Everton Ferreira dos Santos
Eurico E. P. de Lemos

Cambuí (Myrciaria floribunda O. Berg)

Everton Ferreira dos Santos
Eurico E. P. de Lemos

Cambuí (Myrciaria floribunda O. Berg)

Phenology and morphology of plants and post-harvest quality of fruit from cambuizeiro accessions native to Alagoas

ScienciaScripts

Imprint

Cover image: www.ingimage.com

This book is a translation from the original published under ISBN 978-613-9-65620-2.

Publisher:
Sciencia Scripts
is a trademark of
Dodo Books Indian Ocean Ltd. and OmniScriptum S.R.L publishing group

120 High Road, East Finchley, London, N2 9ED, United Kingdom
Str. Armeneasca 28/1, office 1, Chisinau MD-2012, Republic of Moldova, Europe
Printed at: see last page
ISBN: 978-620-7-79425-6

SUMMARY

To my almighty God, for everything you have done in my life and for the opportunities you have given me.

To my Blessed Mother, who protects me and is always guiding my steps.

To my dear and beloved parents:

Ednaldo Alves Ferreira and Tânia Mendes dos Santos;

For love, responsibility and for believing in my potential.

DEDICATE

To my beloved daughter: Marilia Santos Batista The most beautiful, the perfect jewel that God has given me.

To my wife: Thaliny Batista da Silva Santos For her love, patience and motivation.

To my brother:

Ermesson Ferreira dos Santos

For their companionship, friendship and motivation.

OFFER

ACKNOWLEDGMENTS

First of all, I thank God for giving me life, for giving me the opportunity to come so far, for guiding my paths, for giving me the strength to face the difficulties encountered on this journey and for giving me the discernment to always make the right decisions in my life;

To my advisor, Prof. Dr. Eurico Eduardo Pinto de Lemos, for the opportunity, friendship, trust, dedication, for the example of a serious and committed professional and for all his teachings;

To Prof. Dr. Leila de Paula Rezende for her support, motivation, friendship and teachings;

To Prof. Dr. Rychardson Rocha de Araùjo (UFSE), for sharing his knowledge in the area of fruit post-harvest, and for his essential tips and suggestions, which were necessary for the completion of this study;

To Prof. Dr. João Gomes da Costa for his help, support, suggestions and guidance in the statistical analysis of the data;

To everyone in my family for their presence and encouragement throughout my years, which have been fundamental to the consolidation of my values and ethics;

To CNPq for the scholarship;

The coordination and collegiate body of the Postgraduate Course in Agronomy, for all the support during this period of study;

To the entire teaching staff of the Agricultural Sciences Center of the Federal University of Alagoas;

To Tatiana de Lima Salvado, Taciana de Lima Salvador, Cibele Merched Gallo and Jonathan de Araùjo Moraes Ferreira, for their friendship, companionship, for always helping and supporting me when I needed it, as well as for their collaboration in carrying out this work;

I would also like to thank my friends Itawă Gomes Lima de Vasconcelos and Alisson Arquilino da Silva, whom I have gained along the way, from whom I have learned a lot and who have always extended a helping hand when I needed it.

To my friends at the Plant Biotechnology Laboratory (BIOVEG), for their companionship and help in collecting data in the field;

To my master's degree friends: Ivanildo Claudino da Silva, Lekson Santos Rodrigues, Dailson da Silva Oliveira and Renato Carvalho, for their friendship, suggestions and the great times we shared;

To the technicians at the Plant Biotechnology Laboratory (BIOVEG), Gilvânia Moreira da Silva and Hélio do Carmo Lima, for their friendship, support, advice and help with laboratory routines;

Finally, I would like to thank everyone who directly or indirectly contributed to this work.

THANK YOU VERY MUCH!!!

"Sometimes happiness takes a long time to arrive

That's when we can't stop dreaming

Warrior doesn't run away from the fight and can't run away

No one can hold back those born to win..."

"It's a sunny day, but the weather can close in The rain only comes when it has to get wet In life you have to learn If you reap the good you plant It's God who points out the star that has to shine..."

Revelation Group

GENERAL SUMMARY

SANTOS, Everton Ferreira dos. **Phenological and Morphological Characterization of Plants and Postharvest Quality of Fruits of Cambuizeiro (*Myrciaria floribunda* O. Berg) Accessions from the Active Germplasm Bank of CECA/UFAL.** 2018. 121 f. Dissertation (Master's Degree in Agronomy: Plant Production) - Center for Agricultural Sciences/Federal University of Alagoas - CECA/UFAL, Rio Largo - AL, 2018.

The species *Myrciaria floribunda* (West ex Willdenow) O. Berg, Myrtaceae, popularly known as cambui, has a natural distribution in areas of sandy restingas on the south coast of Alagoas, mainly in the municipalities of Piaçabuçu and Penedo. It is a native fruit tree with potential economic value, and its fruit is highly prized both *fresh* and processed. The plant is used as a medicinal and ornamental plant and in urban afforestation projects. However, due to disorderly urban and agricultural occupation, the species is at considerable risk of becoming extinct in Alagoas, necessitating studies aimed at maintaining its genetic diversity in the state. The aim of this study was to evaluate the phenological and morphological aspects of the plants and to carry out the physical and physico-chemical characterization of the fruits of cambuizeiro accessions from the CECA/UFAL Active Germplasm Bank. In the phenological and morphological study carried out at the BAG-Cambui of the CECA/UFAL, in a population of 195 accessions, the leaf change, flowering and fruiting phenophases were observed using a semi-quantitative class scale proposed by Fournier (1974) and the activity index. In the morphological characterization, some vegetative and botanical morphological descriptors of the population of accessions were obtained. The physical and physico-chemical characterization was carried out on 11 accessions taken at random, in terms of fruit weight, seed weight, longitudinal and transverse fruit diameter, seed diameter, total solids (ST), total acidity (AT), ST/AT ratio, vitamin C, carotenoids, anthocyanins, flavonoids, total and soluble pectin. The results were subjected to descriptive statistical analysis, obtaining the minimum, mean and maximum values, standard error of the mean and coefficient of variation, and the means were compared using the Scott-Knott grouping test at 5% probability. Pearson's correlation coefficient was calculated between the variables, and significance was verified using the *t-test* ($P>0.01$). Leaf change occurred throughout the evaluation period, with greater intensity at the end of the rainy season and the beginning of the dry season. The highest peaks of flowering occurred at the end and beginning of the dry season. Fruiting occurred in the months with the highest rainfall, with peaks in May, August and December. The population showed considerable variability in the morphological descriptors determined, leading to the formation of groups with distinct characteristics between the plants. There was a significant positive correlation between the morphometric parameters and the diameter at ground level (DNS) and plant height (AP). There were significant differences between the accessions in terms of the physical and physico-chemical parameters determined, which also led to the formation of groups with different quality characteristics. There was a correlation between the accessions in terms of the quality attributes evaluated. In general, the accessions evaluated showed variability between the parameters evaluated, supporting the selection of accessions based on a single superior characteristic or concomitantly with others.

Index Terms: Myrtaceae, cambuizeiro, accessions, characterization, genetic diversity.

CHAPTER I

GENERAL INTRODUCTION AND LITERATURE REVIEW

1.0 GENERAL INTRODUCTION

Since Brazil is one of the main centers of tropical fruit diversity in the world, it currently stands out as one of the largest producers of fresh fruit, ranking third in the world. It is also home to one of the largest reserves of native plant species in the world, considered to be important chemical, biological and genetic repositories of untapped technologies (MAPA, 2018; SOUZA et al., 2017).

Plant genetic resources are the basis of agricultural exploitation and agrobiodiversity in all its stages. These, together with water and soil, are fundamental to any type of agriculture, so the genetic variability present in native fruit species represents an important tool for food security for current and future generations, given the growing increase in the world's population (SANTILI, 2009; GIACOMETTI, 1993).

The Northeast region is considered to be one of the main centers of diversity of native fruit trees in Brazil, with species that are well adapted to the region's soil and climate conditions. However, there is a scarcity or even absence of information in the literature related to morphological aspects, production, physiological characteristics and phenological behavior, and knowledge of these attributes is extremely important for the description and characterization of superior genotypes, which makes it possible to include these species in commercial production systems (COSTA et al., 2009).

Fruit trees stand out due to their high economic value, both *in* the *fresh fruit* market and in obtaining raw materials for use in the food industry. With regard to the native species, they occupy a prominent position in the place where they occur, as many of them are important sources of food and sustenance for people with lower purchasing power, as well as being a source of income, given that they are sold at open-air fairs, local markets and roadside stalls. In addition, many species produce many nutritious and juicy fruits, which are considered a potential source of mineral salts and vitamins, as well as compounds with biologically functional properties, contributing to the nutrition of the people who live where they occur (AVIDOS; FERREIRA, 2003; QUEIROZ et al., 1999).

Among the potential native fruit trees in Brazil, those of the Myrtaceae family stand out as one of the most important botanical families in the Brazilian flora, due to their wide geographical distribution, as well as the large number of fruit, medicinal and ornamental species (CRUZ, 2014). They correspond to 979 species, distributed in 23 genera, particularly neotropical, with the exception of *Eugenia,* which is pantropical. In general, the native Brazilian species do not produce hardwoods, but their economic importance is due to the fact that they produce edible fruit with high potential for consumption *in natura* and for processing, since, in general, their fruits have an excellent yield of pulp, texture, aroma and characteristic flavor, as well as being rich in bioactive compounds, which act to protect the human body against oxidative damage caused by free radicals (FONTANA et al., 2014).

Among the species belonging to the Myrtaceae family, the Cambui (*Myrciaria floribunda* O. Berg.) is a species of northeastern flora of potential economic value, for preservation and conservation studies, as well as in the development of technologies that enable the installation of commercial orchards. Native to the state of Alagoas, it occurs naturally in the sandy restingas of the south coast, particularly in the municipalities of Coruripe, Feliz Deserto, Piaçabuçu and Penedo. In terms of its botanical features, it is a medium-sized tree with a trunk characterized by the shedding of laminated plates resembling filter paper, and small, white, cymose inflorescences. The fruits are spherical berries that can reach up to 13 mm in diameter. They are yellow, orange, red and burgundy when ripe and have quality characteristics that make them appreciated both *fresh* and processed in the form of sweets, ice cream, liqueurs, jams, among others (ARAÙJO, 2012; SANTOS, 2010).

With a view to maintaining the genetic diversity of the cambuizeiro in Alagoas, and given the high economic and technological potential of this native fruit tree, the aim of this work was to evaluate the phenological and morphological aspects of the plants and to carry out the physical and physico-chemical characterization of the fruits of accessions of *Myrciaria floribunda* O. Berg from the active germplasm bank of the CECA/UFAL, thus establishing parameters that can be used in the domestication and improvement of the species, contributing to its commercial production.

2.0 LITERATURE review

2.1 Brazilian fruit growing

Brazil is a country of extensive territorial dimensions, with an estimated 8,500,000 km^2 , and a wide variation in climate and soil, which makes it ideal for the economic exploitation of a diversity of fruit species (NETO, 2012). According to the Ministry of Agriculture, Livestock and Supply (2018), Brazil was the world's third largest fruit producer with 43.8 million tons, behind only China and India, together they account for around 45.9% of world fruit production, and their production is basically to supply their domestic markets.

Of Brazil's fruit production, 47% is destined for agro-industrial processing and 53% to supply the fresh fruit market. Fruit growing is an economic activity present in all Brazilian states, however, the largest volumes of production are concentrated in just five states, representing 71% of national production (IBRAF, 2012).

The state of Sao Paulo is Brazil's largest fruit producer, with a production of around 15.69 million tons and the most harvested species are oranges, bananas, lemons, tangerines and watermelons. The second largest producer is the state of Bahia with 4.90 million tons of fresh fruit and the main species of fruit grown in Bahia are oranges, bananas and papaya. Next comes Minas Gerais with an estimated production of 3.17 million tons, and the main crops are oranges, bananas and pineapples. Rio Grande do Norte represents the fourth largest supply of fruit, with 2.71 million tons, the main species produced being grapes, apples, watermelons and oranges. The fifth largest producer of Brazilian fruit is the state of Paranà with a production of around 1.60 million tons, with oranges, bananas and tangerines being the most produced fruit (GAZETA, 2017).

Despite the large volume of fruit production, Brazil is still not considered one of the world's leading exporters, as it only exports 2.5% of what it produces. In 2017, Brazilian exports amounted to 784,000 tons of fruit, amounting to US$ 852 million dollars, however, these production figures leave the country outpaced in relation to other Latin American fruit producing countries, such as Chile with US$ 4 billion and Peru with US$ 2.4 billion, leaving it in 23ª position in the world ranking of exporters (MAPA, 2018). The main fruit species exported from Brazil are papaya, mango, watermelon, grapes and lemons, with the European Union being the main buyer of Brazilian fruit. In 2016, the economic bloc absorbed 83.6% of national production, earning US$ 525.1 million (HF BRASIL, 2017). With regard to the main exporting states, in 2016 Ceará stood out as the largest exporter of fresh, dried and processed fruit, as well as walnuts and chestnuts, totaling 162.14 thousand tons and US$ 207.20 million. The second largest was the state of Rio Grande do Norte, where 188.66 thousand tons of fruit were sent abroad, amounting to US$ 142.81 million. The third and fourth largest exporters were the states of Sao Paulo and Pernambuco, which reached US$132.16 million and US$121.77 million, respectively (GAZETA, 2017).

The area harvested in 2017 was 2.581 million hectares, also taking into account the area occupied for the production of nuts and walnuts. The main fruit crops that occupied the largest areas were: oranges (665.17 ha), cashew nuts (586.52 ha), bananas (475.97 ha) and coconuts (251.66 ha). The highest average yields were obtained by papaya and apple, with 48.33 $kg.ha^{-1}$ and 35.28 $kg.ha^{-1}$, respectively. Orange, lemon, melon and pineapple yields ranged from 25,000 $kg.ha^{-1}$ to 26,000 $kg.ha^{-1}$ (GAZETA, 2017).

In the Northeast, fruit growing is also of substantial economic and social importance, given the significant results achieved in recent years. This performance is favored by the prevailing climatic conditions, such as high luminosity throughout the year, high temperature and relative humidity (VIDAL, 2017). However, fruit-growing in the northeast is more centralized in coastal areas due to the high humidity, as well as in irrigated fruit-growing hubs such as Juazeiro da Bahia and Petrolina in Pernambuco, and Mossoró in Rio Grande do Norte, where they mainly produce mangoes, melons, grapes, bananas and pineapples. In the humid highlands, the activity is still practically unexplored and could become an alternative for the production of temperate climate fruit species (VIDAL, 2017; BUSTAMANTE, 2009).

As a result of Brazil's climatic and territorial conditions, which are favorable to the production of tropical, subtropical and temperate fruits, both native and exotic, Brazil has the capacity to become one of the largest fruit production centers *in the* world, not just for the trade of its by-products. It is a form of agricultural exploitation that is constantly developing, especially in relation to new cultivation options, both due to the search by producers and consumers, who are increasingly looking for new fruit options, thus favoring the expansion of production and the market (GONÇALVES et al., 2007; ARAÙJO, 2012; ANDRADE et al., 2008).

Thus, native fruit species such as *Myrciaria floribunda* O. Berg. could become attractive to the growing fruit-growing market when their nutritional and functional potential is properly characterized and publicized.

2.2 Native fruit species with economic and social potential

Native fruit trees, also known as non-traditional food plants, are plants of inestimable socio-economic value that are not commonly consumed or found in commercial centers. They were found in Brazil even before European colonization, as they were already used by the indigenous people for food and medicinal purposes. In this way, a large part of the names of fruit species originated from the Tupi Guarani, and the only reason they haven't become extinct in Brazilian ecosystems is because they are found in places that make it difficult to practice modern agriculture (SILVA et al., 2015; VOLPATO et al., 2015).

These fruits were essential in the diet and nutrition of the explorers and colonizers of Brazil, serving as a source of vitamins and minerals. In this way, species such as pineapple and guarana were incorporated into the eating habits of the Europeans, both for the sake of survival and because of the rich colors, flavors, aromas and textures of the food plants native to Brazil. However, the advanced European colonization and the strong pressure on the indigenous communities gradually extinguished the old habits and cultures, giving way to local traditions and cultures, leading to the spread of Eurasian plants in Brazilian lands, such as sugar cane, wheat, barley, grapes, apples, quince and figs. In this way, foods that were part of indigenous culture were progressively despised and marginalized, such as ingàs, cajàs, araticuns, araçâs, cambucàs, pequis, jatobàs, among other species (KOHLER et al., 2016; VIEIRA et al., 2006).

Brazil, a country of great continental extinction, diversity and endemism of biological species, has the greatest plant genetic diversity in the world. There are approximately 55,000 cataloged species, which represents a wealth of genetic material with potential for use in agriculture, domestication and improvement of fruit species. However, there are still hundreds of neglected fruit species with potential economic value, especially native ones, so knowledge of these fruit trees opens up prospects for creating new products and technologies for an expanding market (SILVA et al., 2016; ARAÙJO et al., 2009).

Brazil's native fruit trees are species that are well adapted to Brazil's soil and climatic conditions and have economic value for both the domestic and foreign markets, both in terms of fresh fruit and processed fruit. However, in order to exploit the full potential of these species, studies are needed to enable their production, conservation and distribution (VOLPATO et al., 2015).

The Brazilian Northeast is a center of diversity for native tropical fruit species. It is estimated that there are around 100 species of native fruit trees with potential economic value that have already been catalogued. Their main characteristic is that they are a source of energy for the people who live *in* the places where they occur, being consumed *fresh* and in other forms such as frozen pulp, juices, sweets, liqueurs, ice cream, medicines, among other forms. In addition, they are also a source of income for the people and communities who sell them at street markets, roadside stalls and municipal markets. However, due to the degradation of the ecosystems where these species occur, as well as the disorganized expansion of agricultural activity, the genetic erosion of several species has occurred, including irreversible losses of germplasm (JÛNIOR et al., 2014).

With regard to the state of Alagoas, it has an extremely rich and varied flora, with species that have enormous potential for use that have yet to be explored. Its flora currently consists of 150 families, 842 genera and 2,002 species. The families with the greatest richness and diversity are: Fabaceae, Asteraceae, Rubiaceae, Malvaceae, Euphorbiaceae, Myrtaceae, Cyperaceae, Orchidaceae and Poaceae (LYRA-LEMOS et al., 2016).

Among the species that make up the Myrtaceae family, *Myrciaria florinbunda* O. Berg., popularly known as cambui, is a fruit tree endemic to the state of Alagoas, with enormous potential for use, given its varied forms of utilization, whether *in natura* or processed, *in* addition to its medicinal and ornamental uses, this species is at serious risk of becoming extinct *in* Alagoas, due to disorganized urban and agricultural expansion.

2.3 Conservation of plant genetic resources

Since man began to grow his own food, agriculture and domestication have always been two strictly related systems, although they did not occur at the same time. With regard to agriculture, three stages were fundamental to its development, characterized as collection and colonization (spontaneous colonization of herbaceous plants that would later be cultivated in places close to human settlements), harvesting (after the colonization stage, species began to be periodically collected, thus characterizing the domestication of plants) and dissemination and planting (stage where seeds were preserved for later sowing, selection and planting). Thus, this was the milestone of the beginning of agriculture and the processes of plant domestication described in the literature, characterized by the first collections of plant germplasm, the acquisition of propagation materials and their subsequent voluntary dissemination to different places (WALTER et al., 2005).

It is known that there are eight centers of diversity of cultivated plants in the world (Figure 1), defined by the then Russian botanist and geneticist, Nicolai Ivanovich Vavilov, who between the 1910s and 1930s undertook expeditions to 52 countries collecting seeds of wild and cultivated species, with the aim of identifying patterns of geographical distribution among species. Thus, these centers would be in 1) China; 2) India (2a. Indo-Malay); 3) Central Asia; 4) Near East Asia; 6) Abyssinia; 7) South and Central Mexico; 8) South America (8a. Peru, Ecuador and Bolivia, 8b. Chile, 8c. Brazil-Paraguay) (PADUA et al., 2007).

Known for its richness in terms of biodiversity, Brazil is one of the most important centers of genetic diversity of plant species in the world, bringing together a wide variety of plants of both economic importance and potential value for future uses (CRUZ, 2014). According to Carvalho et al. (2009), among the countless plant species used for food and agriculture, fruit trees stand out due to their diversity and potential for use, and Brazil is known as one of the main centers of tropical fruit diversity worldwide (Figure 2).

Figure 1 - Center of diversity of cultivated plants according to Vavilov. 1) China; 2) India (2a. Indo-Malay); 3) Central Asia; 4) Near East; 6) Abyssinian; 7) South Mexican and Central American; 8) South American (8a. Peru, Ecuador and Bolivia, 8b. Chile, 8c. Brazilian-Paraguayan). (Source: PADUA et al., 2007).

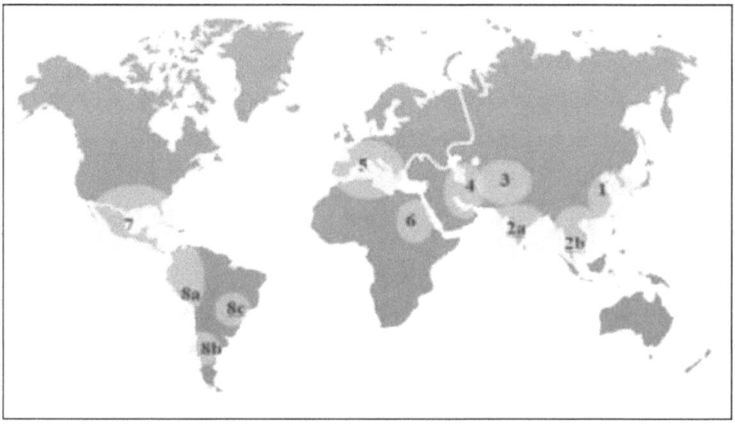

Figure 2 - Centers of diversity of fruit species in Brazil (Source: ARAÙJO, 2009).

However, in Brazil there are still countless unknown or little-known fruit species, in the wild or undomesticated, with the potential to be used for the most diverse purposes, which are neglected by Brazilian research. These species are at risk of disappearing from the ecosystems in which they occur due to the degradation of biodiversity, the expansion of agriculture and livestock, the exploitation of timber resources and also the process of urbanization, which leads to a great loss of natural populations, as well as their genetic diversity, making the conservation and preservation of plant genetic resources increasingly immediate (AGUIAR et al., 2013).

Conservation is defined as the set of policies and programs developed to maintain genetic material in the long term, with potential characteristics for its constant evolution, and preservation is characterized by maintaining individuals or groups of individuals without the consequences of the evolutionary process (CARVALHO et al., 2008). According to Barata et al. (2011), the conservation of genetic resources consists of protecting and maintaining the variability of species with relevant characteristics for current or future uses, in favor of sustainable development. According to the same author, given the current situation in which the world finds itself, in view of climate change and anthropogenic pressure, genetic diversity is increasingly being eroded, and strategies for collection, characterization and conservation are of fundamental importance for maintaining plant genetic wealth.

Phytogenetic resources are understood to be the variability existing between plants or plant populations, elements of biological diversity, of current and potential socio-economic importance, constituting an essential tool for use in plant breeding programs, biotechnology and related areas. They can be conserved inside or outside their natural environment or adapted to both forms, being kept in germplasm banks and collections, "*ex situ*", "*in situ*" or "*on farm*" (JARAMILLO et al., 2000; VALOIS et al., 1996). According to Pàdua et al. (2007), a germplasm bank is defined as:

> The repository where the genetic variability of one or more species is stored. Generally, it consists of the physical base where germplasm is conserved, and can be located in public or private centers or institutions, which conserve germplasm collections in the form of seeds, *in vitro* explants or field plants.

The conservation of plant genetic resources is a task that encompasses many factors, such as obtaining germplasm, characterizing, evaluating, documenting, conserving and using genetic resources (SANTANA et al., 2016). Among the various forms of plant germplasm conservation, fruit species are almost entirely conserved in the field, with few specimens kept in the form of orthodox seeds and/or *in vitro* meristem culture (FERREIRA, 2011). In this way, the success of conservation strategies is achieved by maintaining genetic variability, so knowledge of this is essential (AGUIAR et al., 2013).

Therefore, the characterization of plant germplasm is the starting point for the process of domestication, pre-improvement and improvement of fruit species, as it provides information for the identification of accessions, as well as on the existing variability between accessions (LIMA et al., 2015). According to Carvalho et al. (2002), in the Northeast of Brazil there are a variety of native fruit species of an extractive nature, with a scarcity or absence of information in the literature on morphology, production, physiological and phenological characteristics, making it an extremely important tool for describing and characterizing these fruit trees, as it contributes to their inclusion in commercial production systems, as well as to the conservation of this plant genetic heritage.

There are few Active Germplasm Banks (AGBs) of native fruit trees of the Myrtaceae family in Brazil, among which we can highlight those of the Southern Region, maintained by Embrapa Clima

Temperados in Pelotas-RS. The following species are maintained: guabiroba (*Campomanesia xanthocarpa*); pitanga (*Eugenia uniflora);* araçà (*Psidium cattleianum);* jabuticaba (*Plinia* sp.); cherry (*Eugenia involucrata),* among others (FRANZON et al., 2012).

Furthermore, with a view to preserving the genetic variability of *M. floribunda* O. Berg. in the state of Alagoas, in 2010 approximately 15 fruit collections were made in various locations where the species occurs, precisely in the municipalities of Piaçabuçu and Penedo, which gave rise to the Cambuizeiro Germplasm Active Bank (BAG-Cambui) of the Center for Agricultural Sciences of the Federal University of Alagoas (CECA/UFAL), with 195 accessions that show a certain degree of variability in terms of phenological and morphological aspects and the post-harvest quality of their fruit, which has been the subject of study in various course completion, master's and doctoral studies.

2.4 Family Myrtaceae

The Myrtaceae family is one of the largest botanical families in the world. It is subdivided into two subfamilies: Leptospermae, which has capsular, achene and pixidium fruits, where its dispersion is in Australia and Polynesia, and Myteae, which has dry fruits and its species are widely distributed in South America (BARBIERI et al., 2017; EGUILUZ et al., 2017; ROSARIO et al., 2017).

Comprising around 145 genera and 5,800 species, the representatives of this family occur mainly in tropical and subtropical climate regions, which corresponds to 1.32% of all known Magnoliophyta species in the world. It is estimated that two thirds of these species are found in Brazilian ecosystems (SERAGLIO et al., 2017; TIETBOHl, 2014; VALLILO et al., 2005).

In Brazil, the myrtaceae comprise 23 genera and approximately 1,000 species, making it one of the most economically and ecologically important botanical families in Brazil, as its species are used for food, ornamentation and medicinal purposes, as well as being used in ecological restoration projects, such as the *Eucalyptus grandis* Hill ex Maiden species, which is widely used to restore degraded areas and obtain wood for a wide variety of purposes. In the Northeast, the family ranks eighth in terms of floristic diversity (ARAÙJO, 2012; OLIVEIRA et al., 2016; STADNIK et al., 2015).

In general, Myrtaceae specimens are characterized by being woody plants, shrubs or trees, with entire leaves and alternate or opposite phyllotaxis, and small stipules. Their flowers are usually white or red, ephemeral, hermaphrodite and have radial symmetry. In this family we can find examples of small shrubs that can reach a maximum height of 2 meters, such as the species *Myrcia salzmanni*, as well as large trees over 100 meters high, such as some species of *Eucalyptus* native to Australia (CERQUEIRA, 2002).

There are a large number of fruit species in the Myrtaceae family, some of which are already commercially exploited, such as *guava* (*Psidium guajava* L.), jaboticaba (*Myrciaria cauliflora (*Mart.) O. Berg) and pitangueira (*Eugenia uniflora* L.), either *fresh* or processed into juices, jams, jellies and liqueurs. These species represent a modest portion of the family's high economic potential, given the large number of species that produce edible fruit, but which lack information on their nutritional potential, production and marketing and which, due to the disorganized expansion of agriculture and urbanization, are at serious risk of becoming extinct in the places where they occur (ARAÙJO, 2012; GRESSLER et al., 2006).

2.5 Cambui (*Myrciaria floribunda* O. Berg)

The term "Cambui" is of indigenous origin and means a leaf that comes off. It is a generic name for several specimens included in the *Myrcia* and *Myrciaria* genera. In terms of its botanical aspects, it is a medium-sized tree, characterized by its bark shedding yellowish fragments similar to filter paper. Its inflorescences are white and small. The fruits are small elliptical berries which, when ripe, come in yellow, orange, red and burgundy (Figure 3), and can reach up to 13 mm in diameter (ARAÙJO, 2012).

Figura 3 - General view of the size of the cambuizeiro tree from the Active Germplasm Bank of CECA/UFAL (A), flower buds (B), variability of Cambui fruit colors (C). (Source: author).

Myrciaria floribunda O. Berg has a wide geographical distribution (Figure 4), occurring from Parà to Rio Grande do Sul, in almost all types of plant formations, under the phytogeographic domains of the Amazon, caatinga, cerrado and Atlantic forest biomes, presenting morphological variations (ecotypes), i.e. populations of plants that have adapted to the climate and soil conditions of the places where they occur, becoming exclusive to that environment. The species has several botanical synonyms such as: *Eugenia floribunda* H.West ex Willd, *Myrciaria amazonica* O.Berg., *Myrciaria uliginosa* O.Berg, *Myrciaria leucophloea* O.Berg, *Eugenia salzmannii* Benth, *Myrciaria tenuiramis* O. Berg, Myrciaria *mexicana* Lundel, Myrciaria *oneillii* (Lundell) I.M. Johnst., *Myrciaria protracta* (Steud.) O. Berg. (ARAÙJO, 2012; SANTOS et al., 2014; TIETBOHl, 2014).

Figura 4 - Geographical distribution of the synonyms of the species *Myrciaria floribunda* O. Berg. in the different Brazilian regions (Source: FLORA DO BRASIL (2017), adapted by the author).

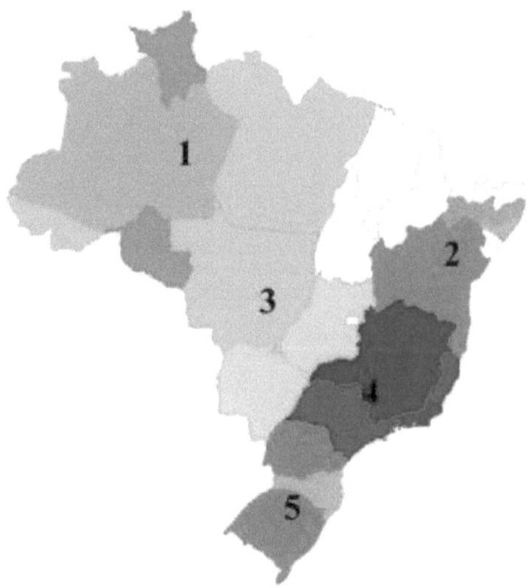

1 - North (Acre, Amazonas, Para, Rondônia, Roraima);

2 - Northeast (Alagoas, Bahia,

Pernambuco);

3 - Midwest (Goias, Mato

Grosso do Sul, Mato Grosso);

4 - Southeast (Espirito Santo, Minas Gerais, Rio de Janeiro, Sao Paulo);

5 - South (Parana, Rio Grande do Sul, Santa Catarina).

The cambui (*M. floribunda* O. Berg.) is a fruit species belonging to the Myrtaceae family with a natural distribution in some restinga areas on the south coast of the state of Alagoas, mainly in the municipalities of Coruripe, Feliz Deserto Piaçabuçu and Penedo. It is a native fruit tree with great potential for commercial exploitation. Its fruit is highly prized both *fresh* and processed, in the form of cakes, ice cream, juices, jellies and liqueurs. The fruit is highly acidic, with a high sugar content, as well as being rich in compounds that have antioxidant action, such as anthocyanins, carotenoids and flavonoids and vitamin C, which can reach 1,526 mg.100^{-1} of pulp (CARVALHO et al., 2012; SANTOS, 2010).

The tree is also widely used in urban afforestation projects, as well as in folk medicine to treat herpes, cold sores, diarrhea and other types of illness (MUNIZ, 2009). Tietbohl et al. (2017), reported that the essential oil extracted from the leaves of *M. floribunda* is rich in phenolic compounds, and especially the flavonoid mycerin-3- galoctoside, compounds that have excellent biological activities, such as anti-microbial, anti-cholinesterase, antioxidant, anti-tumor and insecticide.

It is a non-traditional species of extractive nature, serving as a food supplement and source of income for the populations living in the places where it occurs, as its fruits are sold in roadside stalls, street markets and municipal markets. However, disorganized urban and agricultural occupation represents a considerable risk to the maintenance of its genetic diversity in Alagoas (GAMA et al., 2017; ARAÙJO, 2012). There is a scarcity of information in the literature on the phenological, morphological and biochemical nutritional aspects of the cambuizeiro, and studies are needed to contribute to the domestication and genetic improvement of the species, with a view to developing technologies that enable the establishment of commercial orchards.

2.6 Plant phenology

Phenology is a scientifically and technologically relevant branch of ecology, which investigates biological events associated with environmental characteristics, i.e. the phenological events of individuals that occur periodically under the influence of the biotic and abiotic components of the environment, mainly rainfall and temperature (FRANÇOSO et al., 2014; MARQUES, 2010; ARAÙJO, 2009; FOURNIER et al., 1976).

It is a multidisciplinary science that involves biometeorology, ecology and evolutionary biology, and can be used in applied ecology and conservation biology studies. It characterizes the development cycles of plant species throughout the year, known as phenophases, and is an important tool for understanding the period in which flowers, fruits and seeds are most in supply, as well as their duration, intensity, frequency and synchrony over time, thus enabling the establishment of different phenological behaviours between species, populations and communities of plants that occur in a specific area or region (LUNA-NIEVES et al., 2017; MORELLATO et al.,

2016; PIAUILINO et al., 2012).

Environmental characteristics such as temperature, soil type, photoperiod and rainfall are the main factors that favor the occurrence of plant specimens in certain environments, and these also influence their reproduction time (ASSUNÇÂO et al., 2014). A similar statement was made by Ferrera et al. (2017), who say that the phenological events of plants are influenced by climatic and meteorological variables, such as solar radiation, air temperature, evaporation, precipitation, as well as environmental factors, photoperiod and soil characteristics, in addition to the interaction with biotic elements in the environment, such as pests and diseases.

In tropical regions, the phenological characteristics of plants are always associated with rainfall and soil moisture content, while in temperate regions phenological events are predominantly influenced by day length and temperature (ASSUNÇÂO et al., 2014).

There is a lack of information in the literature on the phenological events of native plant species, especially fruit trees, which makes it difficult to establish standards to be used in conservation and environmental monitoring projects (LUNA-NIEVES et al., 2017). Currently, there is little information on the phenological behavior of *Myrciaria floribunda* O. Berg. in Alagoas, a native fruit species, is scarce in the literature, and the only existing work was carried out by SANTOS (2010), who evaluated the phenology and biometry of cambui fruits from native and cultivated populations in Alagoas, which makes it difficult to collect data to be used in selection, domestication, pre-improvement and improvement work on the species, which is extremely important for setting up commercial orchards and inserting it into the consumer market.

2.7 Morphological characterization of plants

Knowledge of plant morphology is fundamental to taxonomy, as it is used to identify species with a certain degree of phenotypic variation, as well as being important for characterizing a series of physiological events that are manifested as a result of changes in the shape of the plant, thus enabling an understanding of the life history of the specimen studied (ARAÙJO, 2009).

With regard to observing the differences between *already* domesticated plant species, one of the most widely used methods has been morphological characterization using notable and measurable botanical descriptors, with the aim of disseminating information about a given species, contributing to the recognition and identification of plants adapted to a given region, and being essential for describing the genetic diversity of varieties (SALVADOR et al., 2016).

In the characterization of germplasm banks, morphophenological descriptors are one of the most widely used, bringing together easily observed external characteristics of the plant, such as: growth habit, vigour, crown shape, leaf type and shape, flower type, fruiting habit, flowering and fruiting data, fruit size and shape, as well as physicochemical parameters (NUNES et al., 2010). Morphometry is the act of measuring the plant's anatomical structures. It can be carried out using simpler techniques, such as digital calipers and measuring tapes, or more sophisticated ones, such as computerized morphometry. It is worth noting that it does not extinguish the well-established techniques of classical morphology, as well as the qualitative and semi-quantitative analysis used by morphologists. The purpose of this technique is to collect and tabulate information obtained in research quickly and objectively (BELTRÂO et al., 2006).

Information on the morphological aspects of plant species of the Myrtaceae family is relatively scarce in the specialized literature, given the large number of native species, as well as the high forestry and ecological importance of this important botanical family, which lacks studies that unveil its potential (SOARES et al., 2017).

2.8 Fruit quality assessment

The concept of fruit quality encompasses various parameters related to visual appearance, texture, taste and aroma, nutritional value and food safety, thus defining quality (CENSI, 2006). The quality characteristics of horticultural products are organized into categories (sensory, texture, yield, nutritional value and safety), where they must be evaluated as a whole, both to meet consumer demands and to protect public health. In addition to being used as parameters for genetic selection of new varieties, selection of production practices, as well as efficient post-harvest handling practices (MENEZES et al., 2017; CHITARRA & CHITARRA, 2005).

According to Chitarra & Chitarra (2005), quality can be defined as the set of characteristics specific to each horticultural product, which differentiate the individual components and determine the level of consumer acceptance. The quality aspects perceived by the senses are made up of three categories: appearance, texture and taste. Appearance refers to color, brightness, size, shape, integrity, firmness and defects. Texture, on the other hand, encompasses the aspects linked to touch and mouthfeel, such as hardness, softness, juiciness, fibrousness and stickiness. Flavor is related to basic taste, aroma and aftertaste, such as acidic, sweet, bitter, rancid, sweet, astringent and dry (ARAÙJO, 2012).

Quality is the product of the interaction between the physical, chemical and sensory attributes of the fruit, and is an extremely important characteristic for its commercialization, especially when it is destined for the fresh fruit market. The physical and chemical attributes of the fruit are influenced by numerous factors, including: the soil and climate conditions of the growing site, nutritional management, cultivar, crop management, harvest period, among others, so it is important to carefully analyze these parameters in order to verify that they meet the quality standards imposed by the consumer market (AQUINO et al., 2017; ORTEGA et al., 2001).

2.9 Physical and physico-chemical characteristics of fruit.

During the fruit ripening process, characteristic flavors and odors develop in association with an increase in the sugar and acid content. During this process, the fruit softens, accompanied by physical changes, in which chlorophyll is progressively degraded, giving way to carotenoid pigments, flavonoids and anthocyanins. This is the phase in which changes occur in the sensory attributes of the fruit, such as taste, smell, color and texture, which make the fruit acceptable for consumption. Many changes can be assessed both by analyzing visible physical observations and by endogenous changes, such as the content of pigments, acids, sugars, vitamins, tannins, carbohydrates and pectins, among others (CHITARRA; ALVES, 2001).

Thus, with regard to the native fruit tree *M. floribunda* O. Berg, with little information on its physical, physico-chemical and phytochemical attributes, it is important to point out that it could become an attractive option for the Brazilian fruit-growing market, requiring studies on its potential for use, because as well as presenting important economic potential, the information obtained could support the cultivation and appropriate selection of superior genotypes, with a view to their use both for *fresh* consumption and for their use in food industries.

2.9.1 Physical Characteristics

Studies into the physical characterization of fruit provide tools for post-harvest handling and storage, as well as for genetic improvement programs for the species studied. In more advanced stages of commercial exploitation, they also help in the sizing of machinery and equipment in agro-industries, contributing to the appropriate and efficient use of available resources (OLIVEIRA et al., 2009).

From the point of view of the conservation of plant genetic resources, the biometric characterization of fruits and seeds is a valuable tool for assessing genetic variability within and between plant populations, as it allows the differentiation of species of the same genus, to better characterize the influence of environmental factors on the genotypic characteristics of plants, thus allowing the establishment of genetic improvement programs (GUEDES et al., 2015). Furthermore, it is important for understanding and describing germination processes, botanical identification of species, studies related to seed storage and quality tests, as well as seedling dispersal and establishment characteristics, and can be used to differentiate pioneer and non-pioneer species of tropical species (ANDRADE et al., 2010; CÂMARA et al., 2008).

During the fruit ripening process there is an increase in the sugar content in the pulp, where changes in color are usually correlated with sweetness and the promotion of other qualities desirable to consumers, such as an increase in the content of compounds with bioactive properties, such as carotenoids, anthocyanins and flavonoids. The color of a food is essential for its acceptability to consumers, so color is one of the most important quality attributes in fruit, whether fresh or processed (ARAÙJO, 2012).

Color analysis of fruit peel and flesh can be carried out visually using subjective or descriptive scales depending on the degree of ripeness of each product. It can also be carried out using non-destructive chemical or objective methods, where the content of pigments present in the skin of the fruit is assessed, the most commonly used devices for measuring color being spectrophotometers or colorimeters (CHITARRA; ALVES, 2001).

Fruit weight is an important characteristic for marketing *in natura,* as larger fruits are more attractive to consumers. From a technological point of view, fruit with smaller seeds are the most preferred by processing industries because they have higher pulp yields, the minimum value of which processing industries require is 40% (SANTOS et al., 2013).

For fruit to be processed efficiently, plants with the highest fresh mass content must be selected and classified appropriately, as they are the most preferable for industrialization, due to their higher percentage of pulp and higher processing yield. Therefore, the variability in fresh mass and in the yield of its parts influences agro-industrial processes, for example in the manufacture of jams, so efficient classification or separation of the fruit by weight and size must be carried out (REBOUÇAS et al., 2008; VALLILO et al., 2005).

Weight correlates well with the size of the fruit and is a specific characteristic of the variety, so when the fruit reaches its full physiological development, it should show variations in relation to this attribute, and this variation is due to the genotypic characteristics of each variety, as well as climatic influence, and is very flexible (CHITARRA; CHITARRA, 2005).

This characteristic is linearly related to the stage of development and/or ripeness of the fruit, with the exception of fruit that is not ripe enough to be marketed and consumed, as there tends to be a loss of fresh mass due to the higher moisture content and permeability of the exocarp (ARAÙJO, 2012).

Fruit size and shape are quality characteristics that influence consumer choice, harvesting and post-harvest handling, storage capacity, choice of market and final destination of the product. When producing fruit to supply the foreign market, characterizing the size of the fruit is fundamental, given the important requirements of the consumer in relation to this attribute (LIMA, 2015).

It is determined by dimensional characteristics (circumference, diameter, length and width) and weight or volume (specific gravity). It is normally used as a limit for fruit maturity. In terms of shape, this attribute is assessed by the ratio between fruit diameters (DL/DT) or by specific characteristics of the species or cultivar (CHITARRA & CHITARRA, 2005).

The longitudinal (length) and transverse diameters determine the size of the fruit, and their ratio characterizes the shape of the fruit, which is essential for fruit intended for *fresh* consumption, and in some situations for fruit intended for processing, such as in some finished products, where appearance is fundamental, such as jams, candied fruit and glazed fruit (CHITARRA; ALVES, 2001).

The DL/DT ratio indicates the shape of the fruit, so the closer the ratio is to 1, the rounder the fruit is, a characteristic that greatly influences the commercial value of the fruit. Fruits with a ratio close to 1 are preferable in processing industries because they make cleaning and processing easier. Therefore, fruit with a smaller longitudinal diameter than transverse diameter characterizes fruit with a more globular shape (SILVA et al., 2016; SANTOS et al., 2013).

Knowledge of these quality characteristics is important in processing operations, as it facilitates techniques for cutting and peeling raw materials, thus obtaining more uniform products. Fruits with uniform size and weight characteristics are easier to handle in large quantities, since losses are lower, production is faster and quality improves (CHITARRA & CHITARRA, 2005). .

2.9.2 Physico-chemical characteristics

The study, characterization and identification of genetic materials that are productive and, above all, of high quality for *fresh* consumption and industrial use is essential for the formation of commercial orchards. Therefore, in the study of fruit quality, various parameters can be evaluated, which are indicators of the organoleptic characteristics that determine its value as human food (SOUZA et al., 2010; FERREIRA et al., 2006).

. When evaluating fruit quality attributes, various physicochemical indices can be assessed, such as total solids (TS), total acidity (TA), pH, soluble and reducing sugars, starch, total and soluble pectin, total carotenoids, total anthocyanins, yellow flavonoids, polyphenols and total antioxidant activity. According to Júnior et al. (2005), the physico-chemical characteristics of horticultural products can be influenced by various factors, such as genetic makeup, soil and climate conditions, cultivation and post-harvest treatment (SILVA et al., 2015).

In recent years, there has been an increase in the exploitation of products and by-products from certain fruit species, given consumer concern about the association between diet and health. Therefore, information on the chemical composition of fruit is of fundamental importance for understanding its nutritional value, as well as adding value and quality to the final product. With regard to foods with biologically functional properties, these have attracted a lot of attention due to their protective effects on the body against oxidative damage caused by free radicals. Thus, the characterization of compounds that have antioxidant action in fruit is extremely important, as it opens up prospects for their inclusion in the market as a differentiated product (CANUTO et al., 2010).

As the fruit ripens, there is an accumulation of sugars, mainly glucose, fructose and sucrose, accompanied by a decrease in acidity. The sugar content reaches its peak at the end of ripening, giving the product excellent quality. The soluble sugars present in fruit, both in free and conjugated form, are responsible for sweetness, flavor, acid balance, attractive color and texture (CHITARRA; CHITARRA, 2005).

The total solids (TS) content is used as an indirect parameter of the amount of sugar present in fruit, since its concentration tends to increase as it accumulates during the ripening process. It is also used as an indicator of the degree of ripeness in fruit, giving an idea of the level of solids present in the fruit pulp or juice. Its determination does not represent the precise amount of sugars, because other compounds are also dissolved (vitamins, phenolics, pectins, organic acids, among others), but sugars are the most representative, making up 85-90% of total solids (CHITARRA; ALVES, 2001).

According to Pagliarini et al. (2011), the sugar content of fruit is influenced by genotype, environmental characteristics and the management techniques used. This parameter is extremely important in the production of fruit for processing, as it is responsible for a better yield of the processed product. According to Campos (2010), in the agro-industry the total solids content is used as a measure to control the quality of the final product, process control, ingredients and others, such as jams, juices, nectars, pulps, condensed milk, alcohol, sugar, liqueurs and drinks.

The acidity of fruit is due to the organic acids that are dissolved in the vacuoles of the cells, both in free form and conjugated with salts, esters, glycosides, among others. The content of organic acids decreases during the fruit ripening process, as they are used as a substrate in the plant's respiratory metabolism or are converted into sugars, the most abundant being citric, malic, tartaric and acetic, depending on the plant species (CHITARRA & CHITARRA, 2005).

The concentration of organic acids present in food influences taste, odor, color, stability and quality maintenance. It is an extremely important parameter, as it provides relevant information for evaluating the techniques used to process the food, as well as its state of preservation (SOUZA et al., 2010).

Total acidity indicates the acidic or sour taste of fruit and is important not only for indicating the sweetness of a product, but also for its usefulness in food processing industries, such as: product preservation, extending shelf life; basis for ingredient calculations; index for assessing fruit quality and maturity; sensory indicator due to its importance in food flavor and aroma, among other applications (AROUCHA et al., 2010).

The most common method for assessing acidity in fruit is quantitatively by titration, where the amount of acid in a sample reacts with a base of known concentration, using phenolphthalein as a turning point indicator. The results can be expressed in mEq/100 mL or as a percentage of the predominant acid (CHITARRA; ALVES, 2001).

Determining the ratio between total solids content and total acidity in fruit is more representative than evaluating acids or sugars alone, as this is one of the most suitable ways of evaluating fruit flavor (RINALDI et al., 2017).

This ratio can be used to assess the degree of ripeness in fruit, given that total solids increase and organic acids decrease as the fruit grows and matures, making it the best way to assess the degree of ripeness because it is easy to determine and also because it comes close to the actual degree of ripeness. It is normally used as a parameter to describe the taste of the fruit, i.e. it indicates the balance between acids and sugars present in the fruit (ARAÙJO, 2012; SOUSA, 2009).

The term vitamin C refers to a group of compounds that have the biological activity of ascorbic acid. It is the most important nutritional element present in fruit and vegetables, and its content in most products may not exceed 0.3%. There are some fruits that are potential sources of vitamin C, such as acerola (1.0-1.8%), guava and cashew (0.2-0.3%). Citrus fruits do not have high levels of this compound, ranging from 0.04 to 0.08%. However, due to their high daily consumption, they are considered important sources of vitamin C (GIACOBBO et al., 2008; CHITARRA; ALVES, 2001).

It is one of the most important substances for human nutrition because it acts as an antioxidant and is associated with various reactions in the human body, being responsible for the production of collagen, strengthening the immune system, aiding in the absorption of iron and zinc and the elimination of metals, inhibiting nitrosamines and also acting to prevent scurvy (ARAÙJO et al., 2016; SILVA et al., 2009).

Vitamin C is a crystalline substance with an acidic taste and is insoluble in most organic solvents. However, it is soluble in water in the proportion of 1 g to 3 mL. Exposure to heat, air and alkaline environments accelerates its degradation (DANTAS et al., 2012). The human body is unable to synthesize this vitamin because it lacks the enzyme gulonolactone oxidase, which is responsible for the biosynthesis of L-ascorbic acid. Its deficiencies are supplied by food, especially fruit and vegetables (FIORUCCI et al., 2003).

Currently more than 700 different carotenoid molecules have been identified, of which only 40 are present in food, and 13 compounds and 8 metabolites are detected in human tissues. They are pigments derived from tetraterpenes (C_{40}), with a hydrocarbon skeleton of eight isoprene units, and are responsible for the yellow, orange or red color present in many leaves, flowers and fruits, except for the phytoene and phytofluene carotenoids, which are colorless (CHEN et al., 2018; HORST et al., 2009).

In plants, they also have a variety of functions, acting as an attractant for pollinators and seed dispersers, protecting plant organs from damage caused by UV rays, and they are also the compounds responsible for triggering the biosynthesis process of abscisic acid, a phytohormone responsible for regulating a series of plant physiological processes, mainly in the regulation of plant stresses. They also play a biological role in the photosynthetic process, being responsible for the structural stability of both the photochemical reaction center and the antenna complex, among other functions (CHEN et al., 2018; CUONG et al., 2017).

Many factors influence the biosynthesis of carotenoids in vegetables, being influenced by the genotype, growth conditions, stage of development of the vegetable, degree of ripeness, handling during harvesting, processing and storage, geographical area of cultivation of the plant species, environmental conditions and biotic and abiotic stresses (CAYAGO-CRUZ et al., 2018; CAMPOS et al., 2005).

Phenolics are a class of chemical compounds that encompass a diversity of structures, both simple and complex, where they have at least one phenol group, a functional hydroxyl group on an aromatic ring. It is a chemically heterogeneous group of secondary plant metabolites, with around 10,000 compounds already identified. These compounds do not have essential activities in plant development, but they do contribute to the adaptive value of plants, in plant-plant, plant-animal and plant-microorganism relationships. These contributions occur because secondary metabolites are bioactive, as they inhibit herbivory, attract pollinating animals and seed dispersers, protect the plant against ultraviolet radiation, among other functions (TAIZ; ZEIGER, 2013).

Present in greater quantities in vegetables, these compounds have potential health benefits as well as pharmacological properties. In the body, due to their potential to capture free radicals, they act in the prevention and/or reduction of various diseases, such as cardiovascular diseases and various types of cancer, in addition, they have anti-adiabetic activity and also a neuroprotective effect (CHEN et al., 2018).

Polyphenols are phenolic compounds that have an antioxidant action and are the most abundant in the diet. They belong to a large family and are classified according to their structural characteristics into flavonoids (anthocyanins, flavanols, flavonols, flavones, flavavanones, isoflavonoids), phenolic acids, stilbenes, tannins and lignans (YUAN et al., 2017).

Among these compounds, flavonoids are the most widely studied. They have in common the basic structure of 15 carbons, C -C -C$_{636}$ diphenylpropane, divided into two aromatic rings joined by an oxygenated heterocyclic pyran. Around 4,000 flavonoids have been described, the largest groups being flavonols, catechins or flavones, anthocyanidins and isoflavones, where there is a wide range of structural variation depending on the degree of hydrogenation, hydroxylation, methylation and suffonation of the molecules. They also form conjugated complexes with sugar, lipids, amines and carboxylic acids (MEIRA et al., 2016; SUN et al., 2002; DUTHIE et al., 2000).

Flavonoids are polyphenolic compounds that are biosynthesized through two phenylpropanoids and acetate. They contribute significantly to the biosynthesis of various biomolecules, such as aliphatic amino acids, terpenoids, fatty acids, among others. They can be found in various plant species and are of considerable interest to both consumers and scientists due to their numerous therapeutic functions. Potential sources of flavonoids in nature are fruit and vegetables (QUEIROZ et al., 2016; NUNES et al., 2016).

Anthocyanins correspond to a subclass of flavonoids and are glycosides of anthocyanidins, found mainly in flowers and fruits and responsible for most of the blue, purple and countless shades of red. By conferring colors to flowers and fruits, anthocyanins are extremely important as an attractant for pollinators and seed dispersers, in addition to having photoprotective functions, modulating photoinhibition and enhancing the photosynthetic process (YUAN et al., 2017; MEIRA et al., 2016; TAIZ; ZEIGER, 2013; MARKAKIS, 1982).

Within the context of the health-promoting effects of phenolic compounds, numerous studies have described the importance of flavonoids and anthocyanins in protecting and/or inhibiting degenerative diseases (MAGALHÂES et al., 2017). This protective effect has been attributed to the ability of these compounds to capture free radicals present in the human body, thereby reducing the risk of chronic non-communicable diseases. Thus, foods rich in flavonoids and anthocyanins act to regulate the body's physiology through their hypocholesterolemic, hypotensive, atherosclerosis risk-reducing, anticancer, immune system-stimulating and hyperglycemic effects, among others (LOBO et al., 2016). In addition to their antioxidant properties, phenolic compounds have effects on the gastrointestinal tract, binding telomerase inhibitors, regulating signal transduction pathways, inhibiting cyclooxygenase and lipooxygenase, reducing the activities of xanthine oxidase, matrix metalloproteinase and angiotensin-converting enzyme, competing with glucose for transmembrane transport and altering platelet function (MANACH et al., 2005).

Finally, in the face of consumer demand for new products with biologically active properties, native fruit trees stand out as potential sources of antioxidant compounds, which can be used both as functional foods and to provide compounds that can be classified as nutraceuticals, Therefore, the definition of quality standards for the marketing of these fruits for consumption and *in natura* and/or industrialization is extremely important, with a view to promoting human health, as well as making it possible to add value to regionally produced fruits as differentiated products (ANDRADE-WARTHA, 2007).

Polysaccharides are polymers made up of more than 10 monosaccharide units, linked by glycosidic bonds, arranged in a linear or branched form. The most abundant in nature are starch, cellulose, pectins and glycogen. When they contain only one type of monosaccharide, they are called homopolysaccharides or homoglycans, like cellulose, amylose and amylopectin, and when they contain two or more types of monosaccharide, they are heteropolysaccharides or heteroglycans, like alginates. They have the function of retaining water molecules, forming colloidal mixtures and controlling the water activity of an entire system. They have the ability to form gels or viscous solutions with water and are used as thickeners/emulsifiers and stabilizers by the food industry. Pectin, starch, agar and alginates are examples of polysaccharides that have the ability to form gels even at low concentrations (FILHO et al., 2011; TAVERNARI et al., 2008; JUNIOR, 2008).

Pectins are acidic polysaccharides, colloidal in nature, with different levels of esterification and varying degrees of solubility, which play a structural role in the cell wall and intracellular tissues of plants. They consist of polymers of D-galacturonic acid, linked by α-1,4 glycosidic bonds, with one part in the form of a methyl ester. Their chains are terminated by 1,2-α-L-rhamnose units, and they also have a group of neutral sugar side chains, such as D-glucose, L-arabinose, D-galactose and D-xylose (LEÂO et al., 2018; FILHO et al., 2016; PIMENTA et al., 2000).

This polymer is important for giving plant tissues firmness, structure and resistance to compression, as well as being responsible for the texture and palatability of fruit and vegetables during growth, ripening, storage and processing, which are important attributes for the acceptability of the product by the consumer. Thus, knowledge of the constituents of the fruit cell wall is of fundamental importance in post-harvest technology, as it defines management strategies during harvest and post-harvest handling, as well as the marketing destination of the harvested product, i.e. for the fresh or processed fruit market, thus establishing the procedures for the industrial processing of the vegetable product (jam, cut jam, liqueurs, ice cream, jams, among others), the techniques for stimulating the uniform ripening of the fruit and the definition of production flowcharts used in processing industries (SANTI et al., 2014; GONÇALVES et al., 2006).

Their importance in food technology is due to their ability to form gels, being widely used in the manufacture of gums, jellies, dairy products, among others, giving firmness, flavor and aroma to foods. Pectins are usually extracted by chemical and enzymatic methods, consisting of a procedure involving several physical and chemical steps, where hydrolysis, extraction and solubilization are influenced by various factors, including temperature, pH, type of acid and extraction time. Potential sources of commercial pectin are citrus fruits and apple pomace, which are commonly used in the food industry (SEYFRIED et al., 2016; FILHO et al., 2016; MUNHOZ et al., 2008).

In addition to their applications in the food industry, several studies have shown that pectins have biologically functional properties that are beneficial to human health, including lowering serum cholesterol levels, preventing cardiovascular diseases, inactivating heavy metals, protecting against hypertension and diabetes, and also acting to prevent cancer (MOREIRA et al., 2012). According to Souza et al. (2016), it is also currently being used as a soluble dietary fiber due to the physiological effects it has on the human body, reducing levels of cholesterol, bile acids and glucose. Still within the context of the health-promoting properties of pectins, research shows that they are involved in increasing the viscosity of digestive liquor, the thickness of the inner intestinal wall layer, have a protective effect against atherosclerosis and reduce body weight (CANTERI et al., 2012).

Filho et al. (2014) report that the Brazilian flora is rich in fruit species with high biotechnological potential, which require studies aimed at their proper characterization and use. Thus, *M. floribunda* O. Berg, an endemic species from the state of Alagoas, popularly known as cambui, is currently being researched because of its potential to enter the consumer market, given its many uses. In this way, pectins have different properties and consequently different functional properties, making it possible to study alternative sources of extraction and characterization, so the cambui tree could become a new commercial pectin alternative for the food industry.

REFERENCES

AROUCHA, E. M. M., et al. S. Acidez em frutas e hortaliças. **Revista Verde de Agroecologia e Desenvolvimento Sustentàvel**, Mossoró, v.5, n.2, p. 01- 04. 2010.

AQUINO, C. F.; SALOMÂO, L. C. C.; SECON, P. R.; SIQUEIRA, D. L.; RIBEIRO, S. M.

R. Physical, chemical and morphological characteristics of banana cultivars depending on ripening stages. **Revista Caatinga,** v. 30, n. 1, p. 87-96, 2017.

ASSUNÇÂO, V. A.; CASAGRANDE, J. C.; SANTORI, A. L. B. Floristics and Reproductive Phenology of Trees and Bushes in Central West Brazil. **Annals of the Brazilian Academy of Sciences**, v. 86, n. 2, p. 785-799, 2014.

AGUIAR, R. V.; CANSIAN, R. L.; KUBIAK, G. B.; SLAVIERO, L. B.; TOMAZONI, T. A.; BUDKES, J. C.; MOSSI, A. J. Genetic variability of Eugenia uniflora L. in forest remnants at different successional stages. **Revista Ceres**, v. 60, n. 2, p. 226-233, 2013.

ARAÙJO, R. R. **Quality and utilization potential of fruit from Cambui, Guajiru and maçaranduba genotypes native to the coastal vegetation of Alagoas**, Mossoró, 175p., Tese

(Doutorado), 2012.

ANDRADE-WARTHA, E. R. S. **Antioxidant properties of cashew (*Anacardium occidentale* L.) clones: effect on lipoperoxidation and enzymes participating in the antioxidant defense system of the animal organism**. 2007. 111p. Thesis (PhD) - Faculty of Pharmaceutical Sciences, University of Sao Paulo, Sao Paulo, 2007.

ANDRADE, L. A.; BRUNO, L. R. A.; OLIVEIRA, L. S. B.; SILVA, H. T. F. Biometric aspects of fruits and seeds, degree of humidity and overcoming dormancy of jatobà. **Acta Scientiarum**, v. 32, n. 2, p. 293-299, 2010.

AVIDOS, M. F. D.; FERREIRA, L. T. Frutos dos Cerrados - Preservation generates many fruits. In: **Revista Biotecnologia Ciência & Desenvolvimento**, p. 36-41, 2003.

ARAÙJO, R. R. **Phenology and morphology of plants and biometry of fruits and seeds of muricizeiro (*Byrsonima verbascifolia* L. Dc.) from the Tabuleiro Costeiro of Alagoas**. 2009. 81f. Dissertation (Master's Degree in Plant Science) - Universidade Federal Rural do Semiàrido (UFERSA), Mossoró-RN, 2009.

ARAÙJO, R. R.; SANTOS, E. D.; LEMOS, E. E. P.; ALVES, R. E. Biometric Characterization of Fruits and Seeds of Murici Genotypes (*Byrsonima verbascifolia* (L.) Rich.) from the Tabuleiro Costeiro of Alagoas. **Revista Caatinga**, v. 22, n. 3, p. 220-224, 2009.

ANDRADE, R. A.; LEMOS, E. G. M.; MARTINS, A. B. G.; PAULA, R. C.; JÙNIOR, J. L. P. Morphological and Chemical Characterization of Rambutan Fruits. **Revista Brasileira de Fruticultura,** v. 30, n. 4, p. 958-963, 2008.

BARBIERI, S. F.; RUTHES, A. C.; PETKOWICZ, C. L. O.; GODOY, R. C. B.; SASSAKI, G. L.; FILHO, A. P. S.; SILVEIRA, J. L. M. Extraction, purification and structural characterization of agalactoglucomannan from the gabiroba fruit (Campomanesiaxanthocarpa Berg), Myrtaceae family. **Carbohydrate Polymers**, v. 174, p. 887-895, 2017.

BUSTAMANTE, M. A. C. Fruticultura no Brasil e no Vale do Sao Franscisco: vantagens e desafios. **Revista Econômica do Nordeste**, n. 1, v. 40, 20 p., 2009.

BARATA, A. M.; ROCHA, F.; REIS, A.; LOPES, V. R. **The Portuguese Plant Germplasm Bank and the conservation of genetic resources in Portugal**. In Agrorrural: Contributos Cientificos, Instituto Nacional de Recursos Biológicos, I. P. and Imprensa Nacional- Casa da Moeda, S. A., edited by Paula S. Coelho, Pedro Reis. Chapter V - Ecophysiology, genetic resources and plant breeding, pp. 964-974, 2011.

BELTRÂO, F. A.S.; FELIX, L. P.; SILVA, D. S.; BELTRÂO, A. E. S.; LAMOCA- ZARATE, R.

M. Morphometry of accessions of maniçoba (Manihot pseudoglaziovii Pax & Hoffm.) and two species of forage interest. **Revista Caatinga**, v. 19, n. 2, p. 103111, 2006.

CÂMARA, C. A.; NETO, J. C. A.; FERREIRA, V. M.; ALVES, E. U.; MOURA, F. B. P. Morphometric Characterization of Fruits and Seeds and Effect of Temperature on the Germination of Parkia pendula (Willd.) Benth. Ex Walp. **Ciência Florestal**, v. 18, n. 3, p. 281-290, 2008.

CANUTO, G. A. B.; XAVIER, A. A. O.; NEVES, L. C.; BENASSI, M. T. Phytochemical characterization of Amazonian fruit pulps and its correlation with anti-free radical activity. **Revista Brasileira de Fruticultura**, v. 32, n. 4, p. 1196-1205, 2010.

CHEN, Y.; XIE, R.; YANG, J.; CHEN, J.; SUN, Z.; Identification of microbial carotenoids and isoprenoid quinones from Rhodococcus sp. B7740 and its stability in the presence of iron is model gastric conditions. **Food Chemistry**, v. 240, p. 204-211, 2018.

CAYAGO-CRUZ, E.; CORELL, M.; MARIANA, A.; HERNANZ, D.; BENiTEZ- GONZALEZ, A. M.; STINCO, C. M.; MELÉNDEZ-MARTÌNEZ, A. J. Antioxidants (carotenoids and phenolics) profile of cherry tomatoes as influenced by deficit irrigation, ripening and cluster. **Food Chemistry**, v. 240, p. 870-884, 2018.

CUONG, D. M.; ARASU, M. V.; JEON, J.; PARK, Y. J.; KWON, S. J.; AL-DHABI, N. A.; PARK, S. U. Medically important carotenoids from Mormodica charantia and their gene expressions in different organs. **Saudi Journal of Biological Sciences**, v. 24, p. 1913-1919, 2017.

CRUZ, E. S. **Characterization of Jaboticaba Genotypes Based on Morphoagronomic and Molecular Descriptors**. 2014, *76f.* Dissertation (Master's Degree in Resources Plant Genetics) - Federal University of Recôncavo da Bahia (UFRBA), Cruz das Almas - BA, 2014.

CANTERI, M. H. G.; MORENO, L.; WOSIACKI, G.; SCHEER, A. P. Pectin: from raw material to final product. **Revista Polimeros**, v. 22, n. 2. p. 149-157, 2012.

CARVALHO, A. S.; DUARTE, O. R.; VALE, H.S.; SILVA, M. R.; PASSOS, M. A. B. Occurrence and Geographical Distribution of *Myrciaria floribunda* in Roraima. **In: II Brazilian Congress of Genetic Resources,** Belém-PA, 2012.

CARVALHO, C. A. L.; DANTAS, A. C. V. L.; PEREIRA, F. A. C.; SOARES, A. C. F.; FILHO, J. F. M.; OLIVEIRA, G. J. C. **Tópicos em Ciências Agrârias.** Federal University of Recôncavo da Bahia - Cruz das Almas/BA, v.1, 296 p., 2009.

COSTA, M. A. P. C.; SOUZA, F. V. D.; LUNA, J. V. U.; CASTELLEN, M. S.; ALMEIDA, W. A. B.; SILVA, S. A.; DANTAS, A. C. V. L. **Conservation of Potential Fruit Trees for the Brazilian Northeast. Topics in Agricultural Sciences**, v. 1, UFRB, 2009.

CARVALHO, J. M. F. C.; ARAÙJO, S. S.; SILVA, M. A. **Preservation and Exchange of Germplasm.** Embrapa Algodao - Campina Grande/PB, 24p., 2008 (Embrapa Algodao. Documentos 196).

CAMPOS, A. V. S. **Physico-chemical characteristics and mineral composition of *Passiflora setàcea* pulp.** Master's dissertation - University of BrasiliaZFaculty of Agronomy and Veterinary Science, 76 p., 2010.

CAMPOS, F. M.; ROSADO, G. P. New conversion factors for provitamin A carotenoids. **Revista Ciência e Tecnologia de Alimentos**, n. 3, v. 25, p. 571-578, 2005. CERQUEIRA, M. D. **Phytochemical study of *Myrcia rotundifolia* (Berg.) Legrand. (Myrtaceae).** Dissertation (Master's Degree in Chemistry) - Federal University of Bahia (UFBA), Salvador-BA, 2002.

CARVALHO, P. C. L.; FILHO, W. S. S.; RITZINGER, R.; CARVALHO, J. A. B. S. Conservation of Tropical Fruit Germplasm with Farmer Participation. **Revista Brasileira de Fruticultura**, v. 24, n. 1, p. 277-281, 2002.

CENCI, S. A. . Good Post-Harvest Practices for Fruit and Vegetables in Family Farming. **In: Fenelon do Nascimento Neto. (Org.). Basic Recommendations for the Application of the**

Good Agricultural and Manufacturing Practices in Family Farming. la ed. Brasilia: Embrapa Informaçao Tecnològica, 2006, v. , p. 67-80.

CHITARRA, M. I. F.; CHITARRA, A. B. **Postharvest of fruits and vegetables: physiology and handling.** Lavras: UFLA, 785 p., 2005.

CHITARRA, A. B.; ALVES, R. E. **Post-harvest Technology for Tropical Fruits**. Fortaleza: Instituto de Desenvolvimento da Fruticultura e Agroindùstria - FRUTAL/ Sindicato dos Produtores de Frutas do Estado do Ceará, 436 p., 2001.

DANTAS, J. I. A.; PONTES, C. A.; LEITE, G. A.; FERNANDES, P. L. O.; FREITAS, W. E. S.; CARVALHO, C. A. C. Biosynthesis of vitamins in fruits and vegetables. **Revista Agropecuària Cientifica no Semiàrido**, v. 8, n. 4, p. 22-37, 2012.

DUTHIE GG, DUTHIE SJ, KYLE JAM, Plant polyphenols in cancer and heart disease: implications as nutritional antioxidants. **Nutrition Research Reviews**, 2000, 13, 79-106.

EGUILUZ, M.; RODRIGUES, N. F.; GUZMAN, F.; YUYAMA, P.; MARGIS, R. The chloroplast genome sequence from Eugenia uniflora, a Myrtaceae from Neotropics. **Plant Syst Evol**, v. 303, p. 1199-1212, 2017.

FLORA OF BRAZIL 2020. Distribution and synonyms of the species *Myrciaria floribunda* O.

Berg in Available at:

http:<//floradobrasil.jbrj.gov.br/reflora/listaBrasil/PrincipalUC/PrincipalUC.do#CondicaoTaxonCP> Accessed on: December 15, 2017.

FERRERA, T. S.; PELISSARO, T. M.; EISINGER, S. M.; RIGHI, E. Z.; BURIOL, G. A. Phenology of Native Tree Species in the Central Region of the State of Rio Grande do Sul. **Ciência Florestal**, v. 27, n. 3, p. 753-766, 2017.

FILHO, W. B. N.; FILHO, A. A. M. Rheological study of pectin extracted from passion fruit peel. **Revista Virtual de Quimica**, v. 8, n. 3, p. 838-855, 2016.

FILHO, W. B. N.; SOUSA, R. C. P; CHAGAS, E. A.; FILHO, A. A. M. **Evaluation of the yield of pectins extracted from the peel of camu-camu fruit from the northern Amazon.** In: XXIII Congresso Brasileiro de Fruticultura, 4 p., Cuiabà-MT, 2014.

FIORUCCI, A. R.; SOARES, M. H. F. B.; CAVALHEIRO, E. T. G. The importance of vitamin C. **Quimica Nova na Escola**, n. 17, 5 p., 2003.

FRANÇOSO, R.; GUARALDO, A. C.; PRADA, M.; PAIVA, A. P.; MOTA, E. H.; PINTO, J. R. R. Phenology and Fruit Production of *Caryocar brasiliense* Cambess. and Enterolobium gummiferum (Mart.) J. F. Macbr. under Different Burning Regimes. **Revista Arvore**, v. 38, n. 4, p. 579-590, 2014.

FILHO, A. B. M.; VASCONCELOS, M. A. S. **Produçâo alimentici:!: quimica dos alimentos.** Recife - UFRPE, 78 p., 2011.

FERREIRA, A.O.; AGULHEIRO-SANTOS, A.C.; BERNALTE-GARCIA, M.J.& RIBEIRO G. Quality assessment of the 'Rainha Clàudia verde' plum. Instrumental methods vs. sensory evaluation. **Revista de Ciências Agrârias**, 305-308, 2006.

FRANZON, R. C.; RASEIRA, M. C. B. **Germplasm Banks of Native Fruit Trees in Southern Brazil.** Pelotas - RS: Embrapa Clima Temperado (Folder), 2012.

FONTANA, C.; GASPER, A. L.; SEVEGNANI, L. Rare and common species of Myrtaceae from the Semideciduous Seasonal Flora of Santa Catarina, Brazil. **Rodriguésia,** v. 65, n. 3, p. 767776, 2014.

FOURNIER, L. A.; CHARPANTIER, C. The size of the sample and the frequency of observations in the study of the phenological characteristics of tropical trees. **Turrialba**, v. 25, n. 1, p. 45-48, 1976.

GAMA, D.C.; JESUS, J. B.; OLIVEIRA, F. F.; NASCIMENTO JÛNIOR, J. M.; GOMES, L. J. Cambui (*Myrciaria tenella* (DC.) O. BERG; MYRTACEAE): extractivism and income generation in Ribeira do Pombal-Bahia. **Revista Brasileira de Agroecologia,** v. 12, n. 1, p. 4251, 2017.

GAZETA. **Brazilian Fruit Yearbook.** Santa Cruz do Sul: Editora Gazeta Santa Cruz, 88 p., 2017.

GIACOBBO, C. L.; ZANUZO, M.; CHIM, J.; FACHINELLO, J. C. Evaluation of vitamin C content in different groups of araçâ-comum. **Revista Brasileira Agrociência**, v. 14, n. 1, p. 155-159, 2008.

GONÇALVES, M.A.; VIANNA, M. W. H.; BACHA, M. L. **Frutas Brasileiras: Internacionalizacao e Transporte**. In: VII Symposium on Excellence in Management and Technology, 16 p., 2007.

GRESSLER, E.; PIZO, M. A.; MORELLATO, L. P. C. Pollination and seed dispersal in Brazilian Myrtaceae. **Revista Brasileira de Botânica**, v. 29, n. 4, p. 509-530, 2006.

GONÇALVES, C. A. A.; LIMA, L. C. O.; LOPES, P. S. N.; PRADO, M. E. T. Physical, physico-chemical, enzymatic and cell wall characterization at different stages of fig fruit development. **Ciência e Tecnologia de Alimentos**, v. 26, n. 1, p. 220229, 2006.

GIACOMETTI, D. C. Genetic resources of native fruit trees in Brazil. In: National Symposium on Genetic Resources of Native Fruit Trees, 1993, Cruz das Almas - BA. **Proceedings...** Cruz das Almas: EMBRAPA - CNPMF, p. 13-27, 1993.

GUEDES, M. L.; FERREIRA, P. H.; SANTANA, K. N. O.; PIMENTA, M. A. S.; RIBEIRO, L. M. Fruit Morphology and Productivity of Babassu Palms in Northern Minas Gerais State, Brazil. **Revista Arvore**, v. 39, n. 5, p. 883-892, 2015.

HORTIFRUTI BRASIL. **Especial Frutas.** n. 63, 42 p., 2017.

HORST, M. A.; **MORENO** F. S. **Well-recognized functions of nutrients: carotenoids.** 6ª ed. Sao Paulo: Comité de Nutriçao, 2009.

IBRAF. **Overview of the Fruit Production Chain in 2012 and Projections for 2013**.

Brazilian Fruit Institute, 133 p., 2012.

JÛNIOR, J. F. S.; LEDÓ, A. S.; SILVA, A. V.C.; RAMOS, S. R. R. Genetic Resources of Native and Adapted Fruit Trees in the Northeast: status of germplasm conserved *ex situ* in the region. In: III Brazilian Congress on Genetic Resources, 2014, Santos - SP. **Proceedings...** 4 p., 2014.

JÛNIOR, W. E. F. **Carbohydrates, structure, properties and functions. Quimica Nova Escola,** n. 29, 6 p., 2008.

JÛNIOR, J. S. L.; MUSSER, R. S.; MELO, E. A.; MACIEL, M. I. S.; LEDERMAN, I. E.; SANTOS, V. F. Physical and physico-chemical characterization of caj-a-umbu (*Spondias* sp.) fruits. **Ciência e Tecnologia de Alimentos**, v. 25, n. 4, p. 757-791, 2005.

JARAMILLO, S.; BAENA, M. **Conservación Ex Situ de Recursos Fitogenéticos**. Rome: IPGRI, 2000. 209p.

KOHLER, M.; BRACK, P. Native fruits of Rio Grande do Sul: cultivating and valuing diversity. **Agriculturas Magazine**, v. 13, n. 2, 10 p., 2016.

LEÂO, D. P.; BOTELHO, B. G.; OLIVEIRA, L. S.; FRANCA, A. S. Potential of pequi (Caryocar brasiliense Camb.) peels as sources of highly esterified pectins obtained by microwave assisted extraction. **LWT - Food Science and Technology**, n. 87, p. 575-580, 2018.

LUNA-NIEVES, A.; MEAVE, J. A.; MORELLATO, L. P. C.; IBARRA-MONRiQUEZ, G. Reproductive Phenology of Useful Seasonally Dry Tropical Forest Tress: Guiding patterns for seed collection and plant propagation in nurseries. **Forest Ecology and Management**, n. 393, p. 52-62, 2017.

LOBO, A. C. M.; VELASQUE, L. F. L. Literature review on the therapeutic effects of açai and its importance in food. **Biosaiide.** v. 18, n. 2, 10 p., 2016.

LYRA-LEMOS, R. P.; MOTA, M. C. S.; SILVA, J. W. A.; SILVA, A. C. M.; PRATA, A. P. N. **Flora of Alagoas: State of the art, advances and future perspectives**. In: 67th National Botanical Congress - Vitória/ES, Proceedings, 2016.

LIMA, M. S. S.; DANTAS, A. C. V. L.; FONSECA, A. A. O.; BARROSO, J. P. Fruit Characterization of Selected Genotypes of Umbu-cajazeira (*Spondias* sp.). **Interciência**, v. 40, n. 5, p. 311-316, 2015.

MAPA - Ministry of Agriculture, Livestock and Supply. **MAPA to launch plan to increase exports exports.** Available at: <http://www.agricultura.gov.br/noticias/mapa-vai-lancar-plano-para-aumentar-exportacoes- de-frutas> Accessed on: February 14, 2018.

MAGALHAES, M. D.; MACIEL, A. D.; ORSOLIN, P. C. Anticarcinogenic effect of anthocyanin-type flavonoids present in blackberry (*rubus* spp.), identified using the epithelial tumor clone detection test (*wts*) in Drosophila *melanogaster. Revista de medicina e Saùde de Brasilia*, v. 6, n. 1, p. 5-14, 2017.

MORELLATO, L. P. C.; ALBERTON, B.; ALVARADO, S. T.; BORGES, B.; BUISSON, E.; CAMARGO, M. G. G.; CANCIAN, L. F.; CARSTENSEN, D. W.; ESCOBAR, D. F. E.; LEITE, P. T. P.; MENDONZA, I.; ROCHA, N. M. W. B.;

MESQUITA, G. F.; TORQUILHO, H. S. Use of carotenoids for health promotion. **Perspectives on Science and Technology**, v. 8, n. 2, 28 p., 2016.

MEIRA, N. A. N.; PEREIRA, N. P.; MACIEL, L. F.; OLIVEIRA, D. D.; NASCIMENTO, I. S.; SILVA, R. A. Flavonoids and anthocyanins in *Myrciaria cauliflora* (Jabuticaba) for cosmetic application. **Visao Acadêmica**, n. 3, v. 17, p. 50-64, 2016.

MENEZES, P. H. S.; SOUZA, A. A.; SILVA, E. S.; MEDEIROS, R. D.; BARBOSA, N. C.; SORIA, D. G. Influence of ripening stage on the physicochemical quality of umbu fruits (*Spondias tuberosa*). **Scientia Agropecuaria**, v. 8, n. 1, p. 73 - 78, 2017.

MOREIRA, F. K. V.; MARCONCINI, J. M.; MATTOSO, L. H. C. **Thermoplastic pectins: a new class of functional materials for bioactive packaging**. In: Proceedings of the VI Workshop of the Network of Nanotechnology Applied to Agribusiness - Sao Carlos: Embrapa Instrumentaçao; Fortaleza: Agroindùstria Tropical, 2012.

MARQUES, V. B. **Germination, phenology and estimation of the cost of production of the pitahaya (Hylocereus undatus (Haw.) Britton & Rose).** 2010. 141p. Thesis (PhD in Plant Science: Plant Production) - Federal University of Lavras (UFLA), Lavras-MG, 2010.

MUNIZ, A. V. C. S. **Cambui: a native delight.** Aracajù: Embrapa Tabuleiros Costeiros, 2 p., 2009.

MUNHOZ, C. L.; SANJINEZ, E. J.; JUNIOR, M. S. S. Extraction of pectin from dehydrated guava. **Food Science and Technology**, Campinas, 2008.

MANACH C, WILLIAMSON G, MORAND C, Bioavailability and bioefficacy of polyphenols in humans.II Review of 97 bioavailability studies. **American Journal of Clinical Nutrition,** 2005, 81, 243S-255S.

MARKAKIS, P. Stability of anthocyanins in foods. In: MARKAKIS, P. (Ed.) **Anthocyanins as food colors.** New York: Academic Press, 1982. p. 163-180.

NUNES, C. F.; SETOTAW, T. A.; PASQUAL, M.; CHAGAS, E. A.; SANTOS, E. G.; SANTOS, D. N.; LIMA, C. G. B.; CANÇADO, G. M. A. *Myrciaria dubia*, an Amazonian fruit: population structure and its implications for germplasm conservation and genetic improvement. **Genetics and Molecular Research**, v. 16, n. 1, 12 p., 2016.

NUNES, A. G.; LOURENÇO, A. N.; RABELO, C. J.; NETO, F. N. S.; PAULINO, F. A. P.; FERREIRA, E. O. The flavonoid puerarin protects animals from neuronal damage caused by cerebral ischemia. **Mostra Cientifica em Biomedicina**, v. 1, n. 1, 7 p., 2016.

NETO, S. E. A. **Fruticultura Tropical**. Federal University of Acre - Center for Biological and Nature Sciences, Rio Branco/Acre, 232 p., 2012.

NUNES, A. M.; FACHINELLO, J. C.; RADMONN, E. B.; BIANCHINI, V. J.; SCHWARTZ, E.

Morphological and physical-chemical characteristics of butiazeiros (*Butia capitata*) in the Pelotas region, Brazil. **Interciência**, v. 35, n. 7, p. 500-505, 2010.

OLIVEIRA, T. M.; FERREIRA, R. R.; FILHO, J. T.C.; Comparative Analysis of the Initial Growth of *Eucalyptus grandis* Hill ex Maiden (MYRTACEAE) and Guazuma ulmifolia Lam. (Malvaceae). **In: III Congress of Teaching, Research and Extension of UEG**, Pirenópolis- Goiàs, 2016.

OLIVEIRA, M. E. B.; GUERRA, N. B.; MAIA, A. H. N.; ALVES, R. E.; XAVIER, D. S.; MATOS, N. M. S. Physical characterization of pequi fruits native to Chapada do Araripe - CE. **Revista Brasileira de Fruticultura**, v. 31, n. 4, p. 1196-1201, 2009.

ORTEGA, A. L.; RODÉS, A. M. El analisis sensorial como metodo para evaluar La calidad final de las frutas (I). **Dirección General de Tecnologia Agraria - Informaciones Tecnicas,** n. 94, 8 p., 2001.

PIAUILINO, S. A.C.; BOTREL, R. T.; MANCIN, A. C.; SOUSA, A. M. Phenological study of *Lafoensia replicata* Pohl. in the municipality of Bom Jesus, Piaui. **Scientia Plena,** v. 8, n. 4, p. 1-5, 2012.

PAGLIARINI, C. S.; SILVA, F. S.; PORTO, A. G.; SANTOS, P.; LEITE, A. L. M. P. Physico-chemical evaluation of frozen fruit pulps sold in the mid-north region of Mato Grosso. **ENCICLOPÉDIA BIOSFERA,** Centro Cientifico Conhecer - Goiânia, vol.7, n.13, p. 1391-1398, 2011

PADUA, J. G.; FERREIRA, F. R. **Germplasm Banks of Fruit Species.** Toda Fruta, 5p ., 2007. Available at :< https://www.researchgate.net/publication/242745912 BANCOS DE GERMOPLASMA DE ESPECIES FRUTIFERAS> Accessed on: February 1, 2018.

PIMENTA, C. J.; CHAGAS, S. J. R.; COSTA, L. Pectins and pectinolytic enzymes in coffee (*Coffea aràbica* L.) harvested at four stages of ripeness. **Ciência Agrotecnológica,** v. 24, n. 4, p. 1079-1083, 2000.

QUEIROZ, L. P.; ALMEIDA, H. V.; MONTEIRO, J. A.; SILVA, M. G. V. Determination of the total flavonoid content in *Senna fistula* L. *Senna obtusifolia* I. & B. obtained by different techniques. **In: XXXV Encontro de Iniciaçâo Cientifica**, Encontros Universitàrios da UFC, Fortaleza, v. 1, 1 p., 2016.

QUEIROZ, M. A.; GOEDERT, C. O.; RAMOS, S. R. R. **Recursos Genéticos e Melhoramento de Plantas para o Nordeste Brasileiro.** (on line). Version 1. 0. Petrolina - PE: Embrapa Semiàrido/ Brasilia - DF: Embrapa Recursos Genéticos e Biotecnologia, 1067 p., 1999.

REBOUÇAS, L. R.; GENTIL, D. F. O.; FERREIRA, S. A. N. Caracterizaçao Fisica de Frutos e Sementes de Goiaba-da-costa-rica, produzidos em Manaus, Amazonas. **Revista Brasileira de Fruticultura,** v. 30, n. 2, p. 546-548, 2008.

ROSARIO, A. S.; BAUMGRATZ, J. F. A.; SECCO, R. S. Taxonomic studies of *Myrcia* (*Myrciinae, Myrtaceae*) in Brazil: morphological novelties, circumscriptions, and new records for the Amazon. **Iheringia - Série Botânica**, v. 72, n. 2, p. 165-172, 2017.

RINALDI, M. M.; COSTA, A. M.; FALEIRO, F. G.; JUNQUEIRA, N. T. V. Postharvest preservation of *Passiflora setacea* DC. fruits submitted to different sanitizers and storage temperatures. **Brazilian Journla of Food Thcnology**, v. 20, 12 p., 2017. SOUZA, I. J. O.; ARAÙJO, S.; NEGREIROS, P. S.; FRANÇA, A. R. S.; ROSA, G. S.; NEGREIROS, F. S.; GONÇALVES, R. L. G. Diversity of Brazilian flora in the development of health resources. **Revista Uningà**, v. 31, n. 1, p. 35-39, 2017.

SOARES, I. D.; NOGUEIRA, C.; KUNIYOSHI, J. G. S. Morphological characterization of fruit, seed and seedling of *Psidium rufum* DC. (Myrtaceae). **Iheringia**, Série Botânica, v. 72, n. 2, p. 221-227, 2017..

SEYFRIED, M.; SOLDERA-SILVA, A.; BOVO, F. STEVAN-HANCKE, F. R.; MAURER, J. B. B. ZAWADZKI-BAGGIO, S. F. Pectins from medicinal plants: structural characteristics and immunomodulatory activities. **Brazilian Journal of Medicinal Plants,** v. 18, n. 1, p. 201-214, 2016.

SOUZA, R. N. S.; LUCENA, T. K. P.; TAVARES, D. L.; SILVA, A. A. D. **Initial qualitative and quantitative evaluation of pectin in the peel of yellow passion fruit (*Passiflora edulis* flavicarpa).** In: Proceedings of the 1st International Congress on Diversity in the Semi-Arid Region, 8 p., 2016.

SALVADOR, T. L.; LEMOS, E. E. P.; SALVADOR-LIMA, T.; ALVES, G. D. Morphological variations in pine tree genotypes (*Annona squamosa* L.). **Revista Ciência Agricola,** v. 14, n. 1, p. 7-14, 2016.

SANTANA, J. G. S.; NASCIMENTO, T. S.; COSTA, T. M. B.; RABBANI, A. R. C.; SILVA, A. V. C. Estimation of genetic diversity in a natural population of cambui tree (*Myrciaria tenella* O. Berg) using ISSR markers. **Genetics and Molecular Research**, v. 15, n. 4, 11 p., 2016.

STADNIK, A.; OLIVEIRA, M. I. U.; ROQUE, N. Floristic survey of Myrtaceae in the municipality of Jacobina, Chapada Diamantina, State of Bahia, Brazil. **Hoehnea**, v. 43, n.1, p. 87-97, 2016.

SILVA, M. C. B.; MOREIRA, F. J. C.; TAVARES, M. K. N.; SILVA, K. F. Fruit and seed biometry, chemical analysis and pulp yield of yellow araçà (*Psidium cattleianum* Sabine). **Revista Brasileira de Produtos Agroindustriais**, v. 18, n. 3, p. 313323, 2016.

SILVA, N. C. A.; VIDAL, R.; OGLIARI, J. B. A6-437 **Seed exchange networks as a strategy for conserving agrobiodiversity: A case study of popcorn corn in southern Brazil.** In: V Latin American Congress of Agroecology, La Planta - Argentina, 5 p., 2015.

SILVA, V. L.; COZZOLINO, S. M. F. **Vitamin C (ascorbic acid).** In: COZZOLINO, S. M. F. Biodisponibilidade de nutrientes. 3. ed. Barueri: Manole, p. 253-97, 2009.

SANTOS, L. L. C.; BEZERRA, Y. C. A.; COSTA, T. D.; REZENDE, L. P.; LEMOS, E. E. P. Flowering Phenology and Fruit Biometry of *Myrciaria floribunda* O. Berg (Myrtaceae) Cultivated in the Municipality of Rio Largo-AL. **In: LXV Congresso Nacional de Botànica**, Salvador-Bahia, 2014.

SANTI, L. BERGER, M.; SILVA, W. O. B. Pectinases and pectin: commercial application and biotechnological potential. **Caderno Pedagógico**, v. 11, n. 1, p. 130-139, 2014.

SANTOS, J. S.; OLIVEIRA, V. J. S.; BRITO, N. M. Physical Quality of Roma Fruits from Cruz das Almas and Castro Alves. **Textura**, v. 6, n. 12, p. 111-117, 2013.

SANTOS, E. D. **Phenology and Fruit Biometry of Cambui (*Myrciaria floribunda* O. Berg.) from Native and Cultivated Populations in Alagoas.** 2010. 75f. Dissertation (Master's Degree in Agronomy: Plant Production) - Federal University of Alagoas (UFAL), Maceió-AL, 2010.

SANTOS, M. A. C.; QUEIROZ, M. A.; SANTOS, A. S.; SANTOS, L. C.; CARNEIRO, P. C. S. Genetic Diversity among Araçà Accessions from Different Municipalities in the Semi-Arid Region of Bahia. **Revista Caatinga**, v. 27, n. 2, p. 48-57, 2014.

SOUZA, L. M.; CORREIA, K. C.; SANTOS, A. M. G.; BARRETO, L. P.; BEZERRA- NETO, E. **Comparison of methodologies for analyzing pH and titratable acidity in melon pulp.** X Jornada de Ensino, Pesquisa e extensao - UFRPE: Recife, 3 p., 2010.

SOUSA, P. F. C. **Evaluation of sweet orange trees in terms of fruit quality, ripening period and resistance to *Guignardia citricarpa*.** Tese (Doutorado) - Universidade Estadual Paulista, Jaboticabal, 89 p., 2009.

SUN J.; CHU YF.; WU X.; LIU RH. Antioxidant and antiproliferative activities of common fruits. **J Agric Food Chem,** v.50, p.7449-7454, 2002.

SANTILLI, J. **Agrobiodiversity and Farmers' Rights.** Sao Paulo: Peirópolis, 55 p., 2009.

THIETBOHL, L. A. C.; OLIVEIRA, A. P.; ESTEVES, R. S.; ALBUQUERQUE, R. D. D. G.; FOLLY, D.; MACHADO, F. P.; CORRÊA, A. L.; SANTOS, M. G.; RUIZ, A. L. G.; ROCHA, L.

Antiproliferative activity in tumor cell lines, antioxidant capacity and total phenolic, flavonoid and tannin contents of *Myrciaria floribunda*. **Anais da Academia Brasileira de Ciências**, v. 89, n. 2, p. 1111-1120, 2017.

TIETBOHL, L. A.C. **Chemical and biological study of the plant species *Myrciaria floribunda* (H.West ex Willd.) O. Berg**. Universidade Federal Fluminense, Niterói-RJ, 155p., Dissertaçao (Mestrado), 2014.

TAVERNARI, F. C.; CARVALHO, T. A.; ASSIS, A. P.; LIMA, H. J. D. Soluble non-starch polysaccharide in the diet of pigs and poultry. **Revista Eletrônica Nutritime**, v. 5, n. 5, p. 673689, 2008.

TAIZ, L.; ZEIGER, E. **Plant physiology**. 3 ed. Porto Alegre: ARTMED,719p, 2013.

VIDAL, M. F. Recent Behavior of Fruit Growing in BNB's Area of Activity. **Caderno Setorial ETENE**, v. 2, n. 15, 13 p., 2017.

VALLILO, M. I.; GABERLOTTI, M. L.; OLIVEIRA, E.; LAMARDO, L. C. A. PHYSICAL AND CHEMICAL CHARACTERISTICS OF CAMBUCIZE FRUIT (*Campomanesia phaea*). **Revista Brasileira de Fruticultura**, v. 27, n. 2, p. 241-244, 2005.

VALOIS, A.C.C.; SALOMÂO, A.N; ALLEM, A.C. Glossary of plant genetic resources. Brasilia: Embrapa-SPI, 1996. 62 p.

VIEIRA, R. F.; COSTA, T. S. A.; SILVA, D. B.; FERREIRA, F. R.; SANO, S.M. Frutas **Nativas da Regiao Centro-Oeste do Brasil**. Brasilia: Embrapa Recursos Genéticos e Biotecnologia, 320 p., 2006.

VOLPATO, C.; LONGHI, A.; SPERB, M. **Native fruits: local foods, flavors and special ingredients.** CETAP - Centro Ecológico, Rede Ecovida da Agroecologia and Terra do Futuro, 22 p., 2015.

WALTER, B. M. T.; CAVALCANTi, T. B. **Fundamentos para Coleta de Germoplasma Vegetal**. Brasilia: Embrapa Recursos e Biotecnologia, 778 p., 2005.

YUAN, Q.; ZHAO, L. The Mulberry (Morus alba L.) Fruit□ A Review of Characteristic Components and Health Benefits. **Journal of Agricultural and Food Chemistry**, v. 65, p. 10383-10394, 2017.

CHAPTER II

PHENOLOGY AND PLANT MORPHOLOGY OF CAMBUIZEIRA ACCESSORIES (*Myrciaria floribunda* (West ex Willdenow) O. Berg.) NATIVE TO ALAGOAS

SUMMARY

Cambui is a fruit native to Alagoas, belonging to the Myrtaceae family and occurring on the sandy restingas of the state's southern coast. For commercial production, the characterization of *Myrciaria floribunda* O. Berg plants helps in the process of domestication and makes it possible to select accessions with desirable characteristics. The aim of this study was to characterize the phenology and plant morphology of cambui accessions native to Alagoas, in order to establish parameters for distinguishing between them, in terms of the genetic variability of the population, which can be used in genetic improvement programs for the species. The emission of new leaves was continuous throughout the evaluation period, being more intense at the end of the rainy season and the beginning of the dry season (July and October), always accompanied by the abscission of old leaves. Similar behavior was observed for flowering, where the plants flowered during the dry season, with the highest peaks at the end and beginning of the dry season (April and November), when 59.94% and 53.84% of the individuals were in flower. On the other hand, fruiting occurred in the months with the highest rainfall, with the greatest intensity in May with 87.69% of the individuals bearing fruit. The plant population was highly synchronous in all the phenophases studied, with activity rates above 60%. There was a significant positive correlation between the main morphometric indices and the diameter at ground level (DNS) and total plant height (AT). The morphometric variables showed considerable amplitudes between the cambuizeiro accessions, showing the genetic variability of the population and the cluster analyses led to the formation of distinct groups of similarity in terms of the parameters measured.

Index Terms: native fruit trees, phenophases, morphometry, characterization, germplasm bank.

CHAPTER II

PHENOLOGY AND MORPHOLOGY OF CAMBUIZEIRA ACCESSES (*Myrciaria floribunda* (West ex Willdenow) O. Berg.) NATIVE TO ALAGOAS

ABSTRACT

Cambui is a native fruit of the Alagoan flora, belonging to the family Myrtaceae, occurring in the sandy restingas of the south coast of the State. For commercial production the characterization of Myrciaria floribunda plants O. Berg assists in the process of domestication, and makes it possible to select the accessions with desirable characteristics. The objective of this work was to characterize the phenology and morphology of native cambui access plants in Alagoas, in order to establish parameters of distinction between them, regarding genetic variability of the population, that can be used in breeding programs of the species. The emission of new leaves was continuous throughout the evaluation period, being more intense at the end of the rainy season and beginning of the dry season (July and October), always accompanied by the abscission of old leaves. Similar behavior was observed for flowering, where the plants bloomed in the dry season, with the highest peaks at the end and beginning of the dry season (April and November), where 59.94% and 53.84% of the individuals were in bloom. On the other hand, fruiting occurred in the months of higher rainfall indexes, with the highest intensity in May with 87.69% of fruiting individuals. The plant population was highly synchronous in all studied phenophases, considering that the activity index was higher than 60%. There was a significant and positive correlation of the main morphometric indexes with the diameter at the level of the ground (DNS) and the total height of the plants (TA). The morphometric variables presented considerable amplitudes between the cambuizeiro accesses, evidencing the genetic variability of the population and the cluster analyzes led to the formation of different similarity groups regarding the parameters measured.

Index terms: native fruit tree, phenophases, morphometry, characterization, germplasm bank.

1 - INTRODUCTION

The cambuizeiro (*Myrciaria floribunda* O. Berg) is a species belonging to the Myrtacea family, found from the Amazon to Rio Grande do Sul in various plant formations. It is a native fruit of the Alagoas flora, naturally occurring in sandbank environments on the south coast of Alagoas, where it is exploited *in an* extractive manner both for *fresh* consumption and for the production of sweets, liqueurs, ice cream, among others. However, it is a species that is at serious risk of becoming extinct in Alagoas, due to disorganized urban and agricultural occupation, which has been a considerable threat to the conservation of its genetic diversity in the state (SANTOS et al., 2017; SANTOS et al., 2014; ARAÙJO, 2012).

It is a small tree with a round, leafy, low crown and a cylindrical trunk with a diameter of between 30 and 40 cm, characterized by the release of small fragments resembling filter paper. The leaves have a glabrous petiole, elliptic to ovate-lanceolate, glabrous on both sides, with clearly visible secondary veins. Inflorescences are cymose, white and small, gathered in axillary fascicles, with rounded bracteoles. The fruit is globular, smooth, juicy, yellow-orange, red or burgundy when ripe and up to 13 mm in diameter (TIETBOHL, 2014; ARAÙJO, 2012).

It is a non-traditional species, with a shortage of data in the specialized literature, especially in relation to its phenological and morphological characters, requiring work to evaluate and characterize its genetic variability, which can help identify accessions with productive and/or commercially desirable characteristics, thus expanding knowledge of the genetic base of individuals, important for work on the domestication and genetic improvement of the species (MELETTI et al., 2003).

With regard to the vegetative morphological characters of *M. floribunda* plants, there is considerable genetic variability between individuals in terms of growth habit, plant height, tree shape, stem diameter, crown length and diameter, among other characters. Plant morphology is important for understanding inter-dimensional relationships, the space occupied by each plant, assessing the degree of competition in the population of individuals, as well as allowing inferences to be made about the stability, vitality and productivity of each individual. It is a tool for detecting genetic variability within a population of the same species and its relationship with environmental factors, and is a key step in germplasm improvement and conservation programs (BIANCHINI et al., 2016; ANDRADE et al., 2009; DURLO & DENARDI, 1998).

Plant phenology studies recurring vegetative and reproductive events and their relationship with variations in the biotic and abiotic environment, with climate being the main factor controlling phenological patterns in plants over time. It provides information on the establishment and dynamics of plant populations, on the supply of resources throughout the year, on the periods of vegetative and reproductive growth, as well as being a basic tool for establishing scientific methods that enable the rational exploitation of plant genetic resources, especially in the field of fruit growing (PAZ et al., 2018; STRADIC et al., 2018; CAMILO et al., 2013).

In this context, the aim of this work was to learn about the phenological patterns of flowering, fruiting and new leaf sprouting, as well as the morphometric characters of plants from cambuizeiro accessions native to Alagoas, which are part of the CECA/UFAL germplasm bank, establishing parameters for distinguishing between the accessions, in terms of the genetic variability of the population, which can be used in genetic improvement programs for the species.

2 - MATERIALS AND METHODS

Location and characterization of the study area

The work was carried out at the Cambuizeiro Active Germplasm Bank (BAG- Cambui), located at the Centro de Ciências Agrârias da Universidade Federal de Alagoas (CECA/UFAL), in Rio Largo - Alagoas (latitude 9° 29' 45" S, longitude 35° 49' 54" W, altitude 127 meters). According to Koppen, the region's climate is tropical rainy, with dry summers and an average annual rainfall of 1,150.2 mm, the driest months being November to December and the wettest July to August (SEMARH, 2015). The study area has approximately 1,623 m^2 , with a population density of 195 cambuizeiro accessions, arranged at a spacing of 3.0 x 3.5 m, obtained through 15 fruit collections in various locations where the species occurs, precisely in the municipalities of Piaçabuçu and Penedo.

Information on rainfall and temperature was obtained from the CECA/UFAL weather station, approximately 50 meters from the study area. During the period of phenological evaluation of the population of *M. floribunda* accessions, it was possible to observe two clear climatic seasons, a dry one from September to January and a rainy one from April to August, with May being the month with the highest rainfall (584.70 mm) and November the driest (14.20 mm). The average temperature during the study period was 24.6°C, with a minimum of 17.0°C and a maximum of 34.5°C, with no major fluctuations during the experiment (Figure 1).

Figure 1 - Temperatures observed from March 2017 to January 2018 at the Agricultural Sciences Center of the Federal University of Alagoas (CECA/UFAL), in Rio Largo - Alagoas.

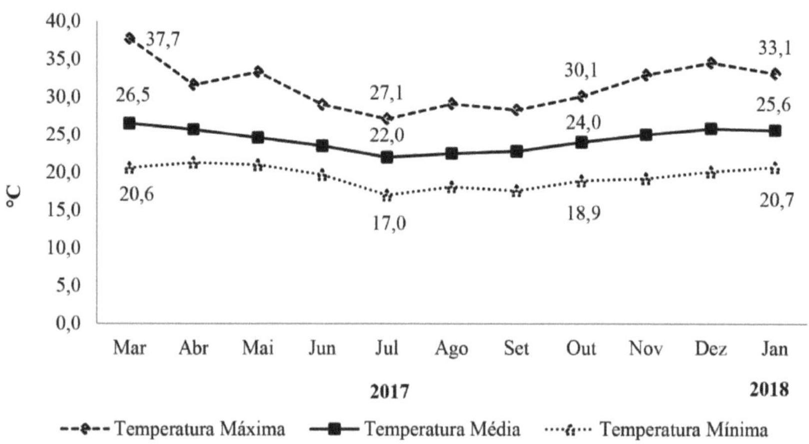

Phenological Characterization

Observations of the flowering, fruiting and leaf emission phenophases were carried out every two weeks from March 2017 to January 2018, to determine the presence or absence of phenological events in the accessions. To analyze the data, phenograms were constructed using two evaluation methods, thus representing the percentage of individuals in the population in a given phenophase per month.

The first evaluation method was the semi-quantitative one proposed by FOURNIER (1974), where the intensity of each phenophase is estimated using an interval class scale of five categories (0 to 4), with 25% magnitude intervals between each one (Table 1). Fournier's intensity percentage was obtained on a monthly basis by adding up the intensity values of all the individuals in each phenophase, dividing the result by the maximum possible value (total number of the population multiplied by four), and the value obtained was multiplied by 100 and expressed as a percentage using the formula below:

Table 1 - Percentage scale of phenological intensity by Fournier (1974), classified into interval classes, with a 25% amplitude between them.

	Fournier scale
Categories	**Description**
0	Absence of phenophase
1	Presence of phenophase varying between 1% and 25%
2	Presence of phenophase varying between 26% and 50%
3	Presence of phenophase ranging from 51% to 75%
4	Presence of phenophase ranging from 76% to 100%

Source: FOURNIER (1974).

$$\text{Percentage of Fournier intensity} = \frac{\Sigma\ Fournier}{4N} \times 100$$

Where:

$\sum$ **Fournier:** sum of the intensity values obtained each month for a given plant species;

N: total number of plants.

The second evaluation method used was the activity index (AI) or percentage of individuals. This method only records the presence or absence of phenophases. It has a quantitative nature, indicating the percentage of individuals in the population that are manifesting a certain phenological event during the evaluation period. It can also estimate the synchrony between individuals in a population (MORELLATO & LEITÂO-FILHO, 1990).

The methodologies used made it possible to evaluate two different patterns of synchronization between the cambium plants (A.I.) and the peaks of flowering and fruiting, as well as leaf emission (Fournier Intensity %).

Vegetative Morphological Characters

The 195 individuals were characterized in terms of morphological indicators: total plant height (AP), measured between ground level and the end of the plant's crown; crown insertion height

(AIC); height of the first bifurcation (APB); crown diameter (DC) which was obtained by measuring four crown radii in fixed directions, north, south, east and west, using a tape measure; and the crown length (CL) was obtained by the difference between the total height of the plant and the height of the crown insertion, all results were expressed in meters; the diameter of the stem at ground level (DNS) was obtained using a digital caliper, and the results were expressed in meters. Based on the morphometric values obtained for the variables described, the following parameters were calculated: overhang index (IS), spread index (IA), crown proportion (PC) and crown shape (FC), according to the method developed by BURGUER (1939) (Figures 2 and 3).

Figura 2 - Quantitative characteristics assessed on the plants of the cambuizeiro accessions, including: total plant height (AP) (A); crown diameter (DC) (B); stem diameter at 10 cm from the ground (DNS) (C and D). Source: author.

Figura 3 - Model of a tree and its morphometric features proposed by BURGUER (1939). Source: CONDÉ et al. (2013), adapted by the author.

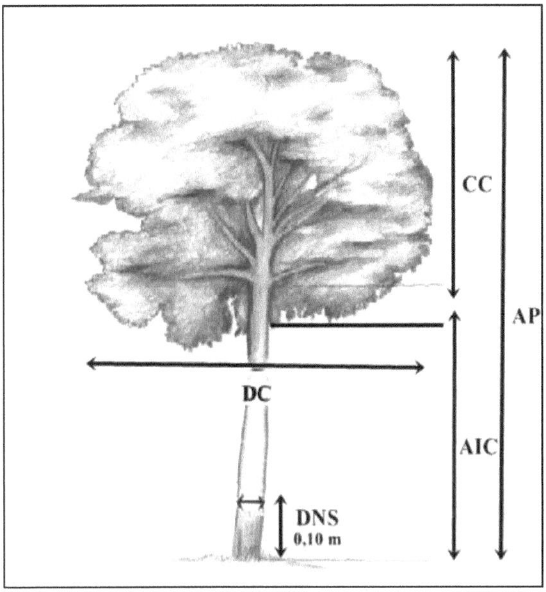

The crown proportion (CP) values were obtained using the equation CP = (CC/AP) x 100, where CC is the crown length in meters and AP is the total height of the plant, also in meters. The overhang index (IS) was determined using the following expression: IS = DC/DNS, where DC is the crown diameter expressed in meters and DNS corresponds to the diameter of the stem at 0.10 m from the ground. With regard to the coverage index, it was obtained by IA = DC/AP, where DC is the crown diameter in meters and AP is the total height of the plant in meters. The crown form was calculated using the equation FC = DC/CC, where DC is the crown diameter and CC the crown length, expressed in meters.

Botanical Morphological Characters

Some botanical morphological characters were also assessed on 50 individuals taken at random, in terms of the number of flower buds, the size of the flower bud, the number of flowers converted into viable fruit, the length of the raceme and the number of nodes per branch. These variables were obtained by counting and measuring with a graduated ruler and digital caliper (6G-150 mm), with the results expressed in centimeters. The results obtained were subjected to descriptive statistical analysis by obtaining the maximum, mean and minimum values, standard error of the mean and coefficient of variation. The accessions were also evaluated for growth habit, obtaining the frequency (%) of individuals with erect and prostrate growth.

Statistical Analysis

The results of the vegetative morphological characters of the plants were subjected to the

Kolmogorov-Smirnov and Lilliefors normality tests to verify the normality and homogeneity of the variances. Subsequently, the data was subjected to descriptive statistical analysis, obtaining the maximum, mean and minimum values and the standard error of the mean. Pearson's correlation coefficient between the variables was calculated, and significance was verified using the *t-test* ($P>0.01$). To assess the genetic variability between the accessions, based on the standardized means of the characteristics studied, the means were compared using the Scott-Knott grouping test at a 5% probability level. All statistical analyses were carried out using the GENES software (CRUZ, 2006).

3 - Results and discussion

Leaf Emission

The change in leaves of the *M. floribunda* accessions was continuous throughout the evaluation period (Figure 4). It was observed that the emission of new leaves occurred simultaneously with the abscission of old leaves, and that the individuals showed considerable leaf fall, but at no time were the canopies completely defoliated. During the study period (March 2017 to January 2018) there were two peaks of greater intensity and frequency of leaf change: 1) at the end of the rainy season, in July, with the population showing an intensity of leaf change of 21.02% and a frequency of 56.92%; 2) and the other at the beginning of the dry season, in October, with an intensity of 42.56% and a frequency of 80.00%; accompanied by variations in temperature and rainfall. In the months following the rainy season, from August 2017 to January 2018, which are the hottest and driest months, a considerable percentage of individuals showed leaf emission (Figures 1, 4).

Figura 4 - Phenological behavior of cambuizeiro accessions native to Alagoas, for the leaf emission phenophase (n=195), based on two analysis methods: percentage of Fournier intensity and activity index.

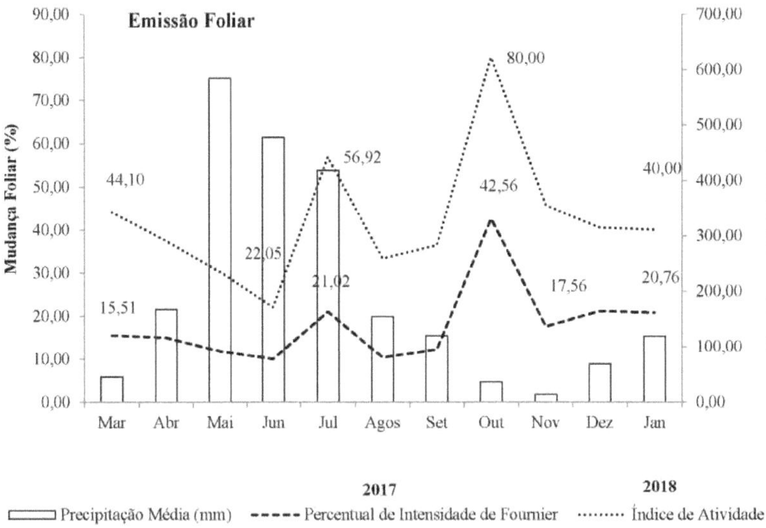

Leaf abscission is a phenomenon influenced by several factors, including the age of the plant, environmental factors, water deficiency and nitrogen deficiency. In this way, the behavior observed in cambuizeiro plants would be a form of vegetative adaptation against the loss of water and carbon, thus allowing individuals to survive in unfavorable conditions, that is, in times of lower water availability (BATALHA et al. 1997). Rainfall is the main abiotic factor that influences the phenological patterns of tropical species, and is most relevant in tropical forests because they have well-defined seasons and notorious seasonality of annual rainfall (ARAÙJO et al. 2014).

During the period of lowest rainfall, leaf fall declined, with 30.25%, 22.05% and 21.02%, respectively. With regard to the intensity of leaf fall, the same behavior was observed, with a decline in the same period, with 15.00%, 11.79% and 10.00% of the individuals showing this phenological pattern.

The population of cambui accessions has a semi-deciduous vegetative pattern, since there is no drastic loss of leaves throughout the year. A similar situation in populations of guabiroba (*Campomanesia xanthocarpa* O. Berg) and pitanga (*Eugenia uniflora* L.) was observed in an area of seasonal semi-deciduous forest in Paranà, as the plants showed leaf change throughout the study period, with the highest percentage of leaf fall in October. With regard to phenological synchrony between the plants, the population showed low synchrony, i.e. an activity index below 25% (AMORIM et al. 2010). A plant population is highly synchronous when the individuals have an activity index equal to or greater than 60%, low synchronous when the individuals have activity between 20% and 60%, and asynchronous when the activity is equal to or less than 20% (BENKE & MORELLATO, 2002).

Flowering

The population of *M. floribunda* accessions flowered in the months with the lowest rainfall. Two flowering peaks were recorded, where the first phase was observed in April, at the end of the dry season, with an intensity of 25.44% and an activity index of 59.94%, and the second in November, at the beginning of the dry season, with an intensity of 22.27% and a frequency of individuals showing this phenological pattern of 53.84%, showing low synchrony in the months of greatest flowering intensity among the accessions (Figure 5).

Figura 5 - Phenological behavior of cambuizeiro accessions native to Alagoas, for the flowering phenophase (n=195), based on two analysis methods: percentage of Fournier intensity and activity index.

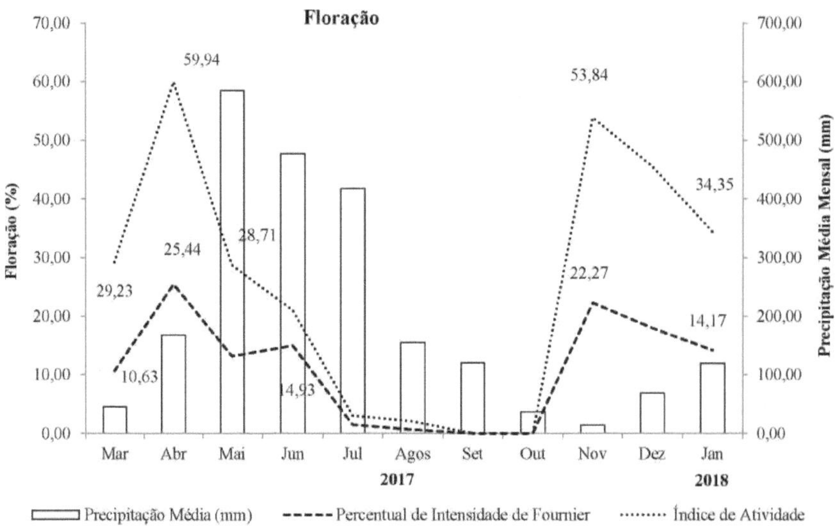

Santos (2010), when evaluating the phenological behavior of native and cultivated populations of cambuizeiro in Alagoas, from June to September 2008 and January to September 2009, observed plants flowering in the dry season and in the rainy season, however, the maximum intensity was recorded in the dry season, with 60% of the population in full bloom, i.e. different results to those obtained in this study. However, in other fruit species of the Myrtaceae family, behavior similar to that observed in the population of cambuizeiro accessions was recorded, such as that obtained by Souza et al. (2008), in which they observed in populations of cagaiteiras (*Eugenia dysenterica* DC.) in the state of Goiás, plants flowering at the end of the dry season and the beginning of the rains.

During the study period, cambuizeiro plants were seen flowering almost all year round, except in September and October, when there was no flowering at all. The lowest flowering intensities occurred in the months with the highest rainfall and mildest temperatures, from May to August, with intensities of 13.16%, 14.93%, 1.51% and 0.63%, and an activity index of 28.71%, 21.02%, 3.07% and 2.05%, respectively.

In natural tableland environments located on the north coast of Alagoas, it is common for species to flower at the end of the rainy season, at the beginning of the dry season. This statement is in line with the behaviour observed among the accessions of *M. floribunda* characterized in this study kept in a germplasm bank in an area of coastal tablelands in the municipality of Rio Largo - AL (ARAÙJO, 2009). According to Muniz (2008), the dry season is considered to be the main external factor controlling the periodicity of the flowering rhythms of tropical vegetation. According to Locatelli & Machado (2004), climate is considered to be the main factor regulating flowering, where through the decomposition of plant remains during the rainy season, the increase in photoperiod and temperature, it is responsible for the entry of energy and nutrients stored in the soil, triggering the flowering process. These factors may explain the phenological flowering behavior of the cambui accessions recorded in this research.

Fruiting

The percentage of individuals with the highest fruiting intensity was recorded in the months of the year when the average monthly temperature was lowest, which is related to the periods of highest rainfall, just after flowering. The fleshy fruits are usually produced in the rainy season, when weather conditions are favorable, making them more attractive and available to frugivores, thus favoring dispersal, germination and seedling growth (MENDES et al., 2012).

Fruiting showed a different phenological pattern among the cambui accessions. Individuals were observed fruiting both during periods of higher and lower rainfall, with no plants fruiting in the months with low rainfall (September to December) (Figure 6). Fruiting began in March with 10.76% of the population bearing fruit, when the average rainfall was 45.7 mm. The highest fruit production occurred in April, in 25.76% of the individuals and an activity index of 57.94%, at the start of the rainy season (584.7 mm). From December onwards, the fruiting intensity of the individuals began to increase, with 23.58% of the individuals fruiting. Fruiting by the species can occur from two to three harvests a year. During the highest peaks in fruit production, there were also high frequencies of fruiting individuals, as the activity rates were higher than 50%

Figura 6 - Phenological behavior of cambuizeiro accessions native to Alagoas, for the fruiting phenophase (n=195), based on two analysis methods: percentage of Fournier intensity and activity index.

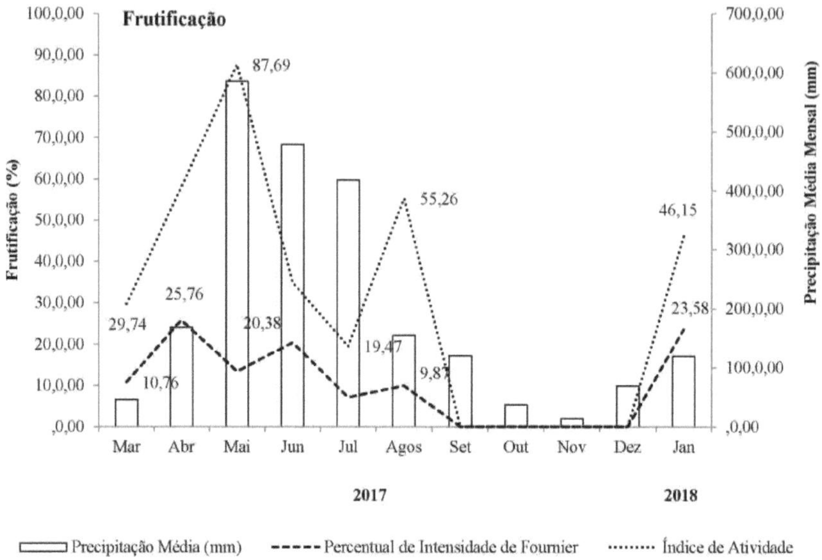

In the phenological study carried out by Santos (2010) on native populations of *M. floribunda*, the highest percentages of individuals in full fruit set were recorded in January and October, with fruit set intensities of 32% and 75%, respectively. According to the same author, there was no fruit production in the rainy season due to the fall of flower buds and flowers caused by the increase in rainfall. It can be seen that the results obtained in this study were different to those reported by the aforementioned author, since the greatest supply of fruit was found during the rainy season.

The phenological behavior of leaf change, flowering and fruiting of the accessions of *M. floribunda accessions* native to Alagoas, should be corroborated by more detailed studies, using data from more years of evaluation, thus verifying whether the phenological characteristics described are reproduced over time, as well as identifying which factors are related to the phenological patterns exhibited, this information being fundamental for the maintenance of the genetic diversity of cambuizeiro in Alagoas, important for the conservation and establishment of individuals in the places where they occur.

Morphology

The results of the botanical characters of the cambuizeiro accessions are shown in Table 2. The average number of floral buds was 43.09 ± 2.36, with a maximum of 76.0 and a minimum of 23.0, indicating that the accessions evaluated express good flowering. The average size of the flower buds ranged from 1.56 mm to 3.18 mm, with an overall average of 2.65 ± 0.05 mm. In camucamuzeiro clones grown in terra firme conditions in Belém - Parà, Marques et al. (2012) found average flower buds ranging from 25.3 to 79.8. The average number of flowers converted into viable fruit ranged from 1.00 to 12.00, with an average of 3.60 ± 0.40 per fruiting branch. The average length of the fruiting branch of the *M. floribunda* species was 28.00 ± 1.43 cm, with ranges from 21.0 cm to 40.0 cm. For the number of nodes per branch, the overall average obtained was 8.58 ± 0.41, with values ranging from 4 to 15 nodes.

Table 2 - Variation and averages of botanical morphological characters of *M. floribunda* plants from the CECA/UFAL Active Germplasm Bank.

Parameters	Maximum	Minimum	Mean ± SD	CV (%)
Number of Flower Buttons	76,00	23,00	43,09 ± 2,36	39,15
Floral Button Size (cm)	3,18	1,56	2,65 ± 0,05	12,26
Number of Flowers Converted into Fruit	12,00	1,00	3,60 ± 0,40	37,82
Fruit Branch Length (cm)	40,00	21,00	28,00 ± 1,43	19,83
Number of nodes/branch	15,00	4,00	8,58 ± 0,41	34,78

EP: standard error of the mean; CV: coefficient of variation.

With regard to plant growth habit, there was no considerable variation between the accessions. The 195 accessions had an erect, more determined and bushy growth habit and 11 (AC27; AC28; AC36; AC52; AC89; AC91; AC111; AC113; AC123; AC167; AC193) accessions grew with well-developed and open branches, characterizing them as prostrate growth (Figure 7).

Figura 7 - Frequency of individuals from the population of cambuizeiro accessions from the CECA/UFAL Active Germplasm Bank, in terms of growth habit.

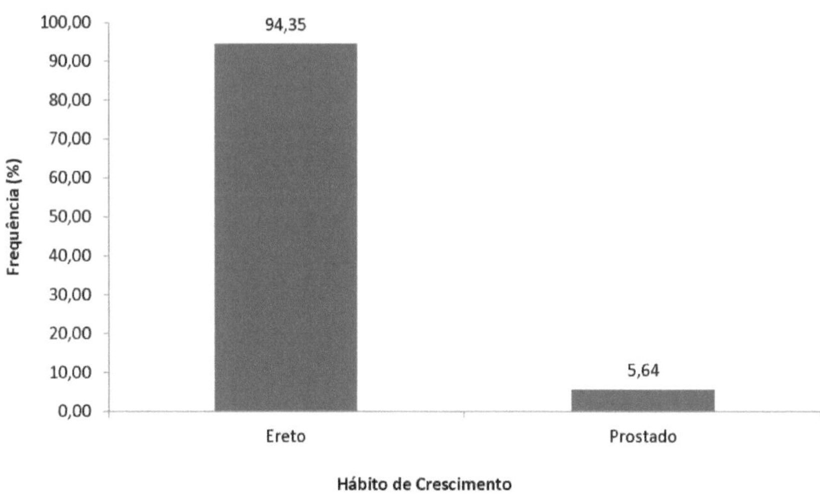

Table 3 shows the morphological characters of the canopy of the population of *M. floribunda* accessions taken for the study, including the minimum, mean and maximum values, the standard error of the mean and the coefficient of variation. Table 4 shows the number of plant groupings for the characteristics evaluated, obtained by comparing the means using the Scott-Knott grouping test at a 5% probability level.

Twenty distinct groups were formed (from a to t) in relation to the total plant height variable, which had an average height of 1.40 ± 0.030 m, with values ranging from 0.40 (AC40) to 2.22 (AC115) meters, and the coefficient of variation obtained was 2.60%. For the morphological characteristic crown insertion height, the ranges observed were from 0.04 m (AC183) to 1.43 m (AC194), with an average point of 0.51 m ± 0.020 and a coefficient of variation of 3.09%. This led to the formation of 23 groups of plants with variations in this characteristic. Siviero et al. (2012) found that when they introduced and evaluated accessions of camu-camu (*Myrciaria dùbia* (H. B. K.) Mcvaugh) in Acre, they found that the total height of the plants varied from 1.14 to 2.62 metres between the accessions.

Table 3 - Variation and means of the vegetative morphological characters of the plants of the accessions of *M. floribunda* from the Active Germplasm Bank of CECA/UFAL.

Characters	Min.	Max.	Mean ± SD	CV (%)
Plant Height (m) (AP)	0,40	2,22	1,40 ± 0,030	2,60
Cup Insertion Height (m) (AIC)	0,04	1,43	0,51 ± 0,020	3,09
Height First Bifurcation (m) (APB)	0,01	1,98	0,42 ± 0,020	4,01
Cup Diameter (m) (CD)	0,25	2,50	1,33 ± 0,024	2,44
Stem Diameter (m) (DNS)	0,008	0,088	0,03 ± 0,005	6,01
Cup length (m) (CC)	0,24	1,87	0,99 ± 0,017	3,93
Cup proportion (%) (PC)	24,31	95,73	70,21 ± 0,729	2,47
Salience Index (IS)	0,48	13,80	4,72 ± 0,074	7,95
index of comprehensiveness (IA)	0,37	1,46	0,95 ± 0,012	5,70
Cup Format (FC)	0,49	2,42	1,38 ± 0,017	4,60

EP - standard error of the mean; CV - coefficient of variation.

Table 4 - Groupings of the morphological characters of cambuizeiro plants, according to the Scott-Knott grouping test at the 5% probability level.

Characters	Number of Groups
Plant Height (m) (AP)	20
Cup Insertion Height (m) (AIC)	23
Height First Bifurcation (m) (APB)	27
Cup Diameter (m) (CD)	24
Stem Diameter (cm) (DNS)	14
Cup length (m) (CC)	17
Cup proportion (%) (PC)	16
Salience Index (IS)	12
index of comprehensiveness (IA)	9
Cup Format (FC)	13

The coefficient of variation obtained for the variable height of the first fork was 4.01%, indicating considerable variability for this characteristic, and 27 groups of plants were formed. Accessions AC96, AC101 and AC105 had the highest average values, i.e. 1.98, 1.75 and 1.69, respectively. The lowest height observed was in accession AC34, with 0.01 meters. On average, the plants were 0.42 ± 0.02 meters in relation to this parameter. Júnior et al. (2009) carried out the genetic characterization of a germplasm collection of caigateira (*Eugenia dysenterica* DC.) native to the cerrado, and found that the height of the first fork ranged from 0.09 to 1.37 metres among the accessions.

The canopy diameter values ranged from 0.25 m (AC40) to 2.50 m (AC106). For this variable, the accessions were grouped into 24 different groups, with an overall average of 1.33 ± 0.024 m and a coefficient of variation of 2.44%. According to Sanquetta et al. (2014), crown diameter can be taken as a basic parameter for estimating other morphometric variables in plants, which corresponds to the distance between the lines of projection of the outermost points of the crown. In the morphological characterization of rambutam (*Nephelium lappaceum* L.) plants carried out by Andrade et al. (2009), it was found that the crown diameter varied from 1.98 and 3.60 metres (in the planting line) to 1.35 and 2.50 metres (between plants). In this study, the crown diameter values were similar to those obtained by the aforementioned authors for rambutan plants.

For the accessions evaluated in the field, the average DNS was 0.30 ± 0.005 cm, with the accessions with the largest diameter being 0.88 cm and the smallest 0.08 cm, corresponding to accessions AC44 and AC40, with 14 groups of plants with different stem diameters being formed. In the study carried out by Pinto et al. (2016), which characterized gabirobeira plants (*Campomanesia adamantium (*Cambess.) O. Berg) from different sources in the state of Goiás, the average stem diameter was 14 mm, with a minimum of 7.48 mm and a maximum of 20.70 mm. In camucamuzeiro clones, Marques et al. (2011) found an average stem diameter of 13.54 cm. The results presented in this study were lower than those obtained for other fruit species in the Myrtaceae family.

The crown length obtained from the difference between total plant height and crown insertion height ranged from a maximum of 1.87 m to a minimum of 0.24 m, with an overall average of 0.99 ± 0.017 m and a coefficient of variation of 3.93%. The longest accessions were AC96, AC101 and AC106, with values of 1.84, 1.78 and 1.67 meters, respectively. With regard to this variable, 17 different groups of plants were formed. The average proportion of canopy exposure (CP) in relation to total plant height was 70.21 %, ranging from 24.31 % (AC44) to 95.73 % (AC150). According to the grouping test, the plants were grouped into 16 classes (from a to p), with the highest canopy proportions being exhibited by accessions AC150, AC167, AC123 and AC183, with values of 93.73, 95.50, 93.30 and 93.25 %, all belonging to group A. In their study of morphometric relationships in *Ocotea odorifera* (Vell.) Rohwer plants, Orellana & Koehler (2008) found an average crown proportion of 55.2%, with a maximum value of 69.23% and a minimum of 30.0%.

The overhang index (IS) is a parameter used to estimate how many times the crown is larger than the DNS. The cambuizeiro shrubs characterized in this study have an average crown diameter 4.52 times larger than the DNS. Twelve distinct groups (from A to L) were obtained between the plants for this parameter, with the maximum value observed being 13.80 (AC161 from group A) and the minimum 0.48 (AC44 from group L). There was high variability between the accessions for this morphometric parameter, given that the coefficient of variation was 7.95%. Durlo (2001) obtained an average salience index for *Cabralea canjerana* (Well.) Mart. of 21.7, with a minimum of 12.8 and a maximum of 35.7. This index is used to control plant spacing in order to reduce competition. In the morphometric characterization of *Jatropha curcas* L. plants, indices ranging from 12.12 to 16.51 were obtained, values higher than those obtained for the plants of the *M. floribunda* accessions measured in this study (FEY et al., 2014).

The average value for the coverage index (CI), which corresponds to the ratio between crown diameter (CD) and total plant height (PH), was 0.95 for the *M. floribunda* plants. The range observed was 0.37 (AC46) to 1.46 (AC123), and the plants were grouped into nine distinct groups (from a to i). The coefficient of variation obtained was 5.70%, indicating considerable variation between the accessions. If there is a correlation between this index and plant height, it can be used as a thinning criterion during the life cycle of the plant stand, providing information on the number of plants that can fit into a hectare without competition between them (ROMAN et al., 2009). In the study of the morphometry and competition of *Trichilia claussenni* C. DC., which occurs in fragments of semi-deciduous forest in Rio Grande do Sul, carried out by Silveira et al. (2015), coverage indices ranging from 0.34 to 1.17 were obtained, with an average of 0.65 ± 0.20. Araùjo (2009) carried out a morphological study of *Byrsonima Vesbascifolia* (L.) Rich. plants in a coastal tableland area of Alagoas, and obtained coverage indices ranging from 0.25 to 0.86. The results obtained from cambui plants native to Alagoas are similar to those found by the authors cited for other plant species.

With regard to the characteristic crown shape (CF), 13 groups were formed (from a to n). The values ranged from 0.49 (AC41) to 2.42 (AC6), and the average point was 1.38 ± 0.017, indicating that among the accessions there are both plants with more slender canopies (CF ≤ 1) and those with rounder, flatter shapes (CF > 1). Tonini & Arco-Verde (2005) carried out a morphometric characterization of four native Amazonian plant species and found values of 0.30, 0.73, 0.79 and 0.86 for the species andiroba (*Carapa guianesis* Aubl.), Brazil nut tree (*Bertholletia excelsa* Bonpl.), ipê-roxo (*Tabebuia avellanedae* Lorentz ex Griseb.) and jatobà (*Hymenaea courbaril* L.), respectively.

Table 5 shows that there is a significant positive correlation between DNS and plant height (AP), crown insertion height (AIC), height of first bifurcation (APB), crown diameter (DC), crown length (DC), crown proportion (PC), overhang index (IA) and crown shape (FC), indicating that these morphometric parameters tend to increase as DNS increases. The negative correlation between the diameter at ground level and the overhang index shows that it tends to decrease as the DNS increases. It was observed that for plant height there was also a positive and significant correlation between the variables, where they tend to increase as plant height increases, except for crown shape, where the correlation was negative and significant, showing that plant shape decreases with increasing plant height. There was also a negative correlation with plant height for the comprehensiveness index, but it was not significant.

The results presented show the genetic variability between the individuals in the population studied, representing an important tool for management strategies and the selection of accessions of this species. In addition, the information is useful for sizing plant spacing and population density in the field, given that the cambuizeiro is not a domesticated fruit species and that commercial production systems do not yet exist.

Table 5 - Pearson's correlation coefficients between diameter at ground height (DNS) and plant height (AP) and the morphometric parameters of *M. floribunda* from the Cambui Active Germplasm Bank at CECA/UFAL.

Variables	AP	AIC	APB	DC	CC	PC	IS	AI	FC
DNS	0,5261**	0,2027**	0,3044**	0,5603**	0,5040**	0,1758**	-0,2885**	0,1883**	0,1607**
AP	---	0,2915**	0,3097**	0,6887**	0,8059**	0,1285**	0,1753**	-0,0207	-0,1661**

** Correlation at a significance level of 0.01.

4 - CONCLUSIONS

1 - The flowering phase followed the seasonal climate, influenced by rising temperatures and low rainfall;

2 - The fruit supply of the species is relatively quick, but there are two to three harvests a year, coinciding with the period of highest rainfall.

3 - The phenological behavior of the population followed the climatic variations,

confirming the influence of abiotic factors on the manifestation of phenological events, especially rainfall;

4 - The morphometry of the canopy varies according to the space occupied by the plant;

5 - When the morphometric characteristics were correlated with the DNS, only the salience index showed a negative and significant correlation, where it tended to decrease as the DNS increased;

6 - In the correlation between morphometric characteristics and plant height (PH), there was a negative and significant correlation only with crown shape (CF), indicating that it decreases with increasing plant height.

REFERENCES

ARAÙJO, R. R. **Quality and utilization potential of fruit from Cambui, Guajiru and maçaranduba genotypes native to the coastal vegetation of Alagoas**, Mossorò, 175p., Tese (Doutorado), 2012.

AMORIM, J.S., AMORIM, J.S., CERQUEIRA-SILVA, C.B.M., ROMAGNOLO, M.B. Phenology of the species *Campomanesia xanthocarpa* O.Berg and *Eugenia uniflora* L. (Myrtaceae) in the municipal forest of Paranavai, Paranà. **Electronic Journal of Biology,** 3: 84-98, 2010.

ANDRADE, R. A.; LEMOS, E. G. M.; MARTINS, A. B. G.; PAULA, R. C. Morphological characterization of rambutan plants. **Maringà,** v. 31, n. 4, p. 613-619, 2009.

ARAÙJO, R. R. **Phenology and morphology of plants and biometry of fruits and seeds of muricizeiro (*Byrsonima verbascifolia* L. Dc.) from the Tabuleiro Costeiro of Alagoas**. 2009. 81f. Dissertation (Master's Degree in Plant Science) - Universidade Federal Rural do Semiàrido

(UFERSA), Mossoró-RN, 2009.

ARAÙJO, R. R.; SANTOS, E. D.; LEMOS, E. E. P. L. Phenology of the muricizeiro Byrsonima verbascifolia (L.) Rich in a coastal tableland area of northeastern Brazil. **Revista Ciência Agricola,** v. 12, n. 1, p. 1-8, 2014.

BIANCHINI, F. G.; BALBI, R. V.; PIO, R.; SILVA, D. F.; PASQUAL, M.; BOAS, E. V. B. V. Morphological and chemical characterization of cambucizeiro fruits. **Revista Bragantia**, v. 75, n. 1, p. 10-18, 2016.

BENCKE, C.S.C.; MORELLATO, L.P.C. A comparative study of the phenology of nine tree species in three types of Atlantic forest in southeastern Brazil. **Revista Brasileira de Botânica**, v. 25, n. 2, p. 237-248, 2002.

BATALHA, M. A.; ARAGAKI, S.; MANTOVANI, W. Phenological variations of cerrado species in Emas (Pirassununga, SP). **Acta. Bot. Bras.**, v. 11, n. 1, 18 p., 1997.

BURGER, H. Baumkrone und zuwachs in zwei hiebsreifen Iichtenbestanden. **Mitteilungen der Schweizerischen Anstalt für das Forstliche Versuchswesen**, v.21, p.147-176, 1939.

CONDÉ, T. M.; LIMA, M. L. M.; LIMA-NETO, E. M.; TONINI, H. Morphometry of four forest species in agroforestry systems in the municipality of Porto Velho, Rondônia. **Revista Agro@mbiente On-line**, v. 7, n. 1, p. 18-27, 2013.

CAMILO, Y. M. V.; SOUZA, E. R. B.; VERA, R.; NAVES, R. V. Phenology, production and precocity of *Eugenia dysenterica* plants for genetic improvement. **Revista de Ciências Agrârias,** v. 36, n. 2, p. 192-198, 2013.

CRUZ, C.D. **GENES program**: multivariate analysis and simulation. Viçosa: UFV, 2006. 175p.

DURLO, M.A. Morphometric relationships for *Cabralea canjerana* (Well.) Mart. **Ciência Florestal,** v. 11, n. 1, p. 141, 149, 2001.

DURLO, M. A; DENARDI, L. Morphometry of *Ocotea odorifera* in native secondary forest of Rio Grande do Sul. **Ciência Florestal**, Santa Maria, v. 8, n. 1, p. 55-66, 1998.

FEY, R.; MALAVASI, U. C.; MALAVASI, M. M.; SCHLZ, D. G.; DRANSKI, J. A. L. Interdimensional relationships and productivity of stone pine (*Jatropha curcas* L.) in a silvipastoral system. **Semina: Ciências Agrârias**, v. 35, n. 2, p. 613-624, 2014.

FOURNIER, L. A. Un mètodo cuantitativo para la medición de caracteristicas fenológicas en árbores. **Turrialba**, v. 24, n. 4, p. 422-423, 1974.

JÛNIOR, E. B. A. **Quantitative and molecular genetic variability in a germplasm collection of**

Eugenia dysenterica DC. (Myrtaceae), Goiânia - GO, 88 p., Dissertation (Master's Degree in Genetics and Plant Breeding), 2012.

JUNIOR, E. B. A.; CHAVES, L. J.; SOARES, T. N. Genetic characterization of a germplasm collection of cagaiteira, a native species of the cerrado. **Bragantia,** v. 68, n. 3, p. 629637, 2009.

JESUS, N. MARTINS, A. B. G.; ALMEIDA, E. J.; LEITE, J. B. V.; GANGA, R. M. D.; JUNIOR, E. J. S.; ANDRADE, R. A.; MOREIRA, R. F. C. Characterization of four groups of jaboticabeira, under the conditions of Jaboticabal-SP. **Revista Brasileira de Fruticultura,** v. 26, n. 3, p. 482-485, 2004.

LOCATELLI, E.; MACHADO, I. C. **Phenology of tree species in a mountain forest (Brejo de Altitude) in Pernambuco, Northeastern Brazil**. In: K. C. Porto; J. J. Cabral; M. Tabarelli (Org.). Brejos de altitude: história natural, ecologia e conservaçao. MMA/PROBIO, Brasilia, p. 255, 2004.

MARQUES, D. N.; OLIVEIRA, M. S. P.; NASCIMENTO, W. M. O. Evaluation of flowering characters in camucamuzeiro clones under terra firme conditions in Belém - Parà. In: Congresso Brasileiro de Recursos Genéticos, 2., 2012, Belém, PA. **Proceedings...** Brasilia, DF: Brazilian Society of Genetic Resources, 2012.

MENDES, F. N.; RÊGO, M. M. C.; ALBUQUERQUE, P. M. C. Phenology and reproductive biology of two species of *Byrsonima* Rich. (Malpighiaceae) in a Cerrado area in Northeastern Brazil. **Biota Neotropical**, v. 11, n. 4, p. 103-115, 2011.

MARQUES, D. N.; OLIVEIRA, M. S. P.; LOPES, V. S.; NASCIMENTO, W. M. O. Vegetative evaluation of camucamuzeiro clones in Belém - Parâ. **In: 15° Seminàrio de Iniciaçâo Cientifica da EMBRAPA,** Embrapa Amazônia Oriental, Belém - PA, 4 p., 2011. MUNIZ, F. H. Flowering and fruiting patterns of trees in the Maranhão Amazon. **Acta Amazônica**, v. 38, n. 4, p. 617-626, 2008.

MELETTI, L. M. M.; BERNACCI, L. C.; SOARES-SCOTT, M. D.; AZEVEDO-FILHO, A.; MARTINS, A. L. M. Genetic variability in morphological, agronomic and cytogenetic characters of sweet passion fruit (*Passiflora alata* Curtis) populations. **Revista Brasileira de Fruticultura,** v. 25, n. 2, p. 275-278, 2003.

MORELLATO, L.P.C.; LEITÂO-FILHO, H.F. Phenological strategies of tree species in mesophytic forest at Serra do Japi, Jundiai, Sao Paulo. 50 (1):163-173. 1990.

ORELLANA, E.; KOEHLER, A. B. Morphometric relationships of *Ocotea odorifera* (Vell.) Rohwer. **Rev. Acad. Agrâr. Ambient.**, Curitiba, v. 6, n. 2, p. 229-237, Apr./Jun. 2008. PINTO, J. F. N.; SANTOS, E. A.; REIS, E. F.; SILVA, D. F. P.; PEREIRA, L. D. Characterization of gabirobeira from different sources in the state of Goiás. **In: V SIMPA- Simpósio da Pôs-Graduaçâo em Agroecologia da Universidade Federal de Viçosa**, p. 4144, Viçosa - MG, 2016.

PAZ, W. S.; SOUZA, J. T. Phenology of arboreal plant species in a caatinga fragment in Santana do Ipanema, AL, Brazil. **Diversitas Journal,** v. 3, n. 1, p. 39-44, 2018. PEREIRA, T. S.; COSTA, M. L. M. N.; MORAES, L. F. D.; LUCHIARI, C. Phenology of tree species in the Atlantic Forest of the Poço das Antas Biological Reserve, Rio de Janeiro, Brazil. **IHERINGIA,** v. 63, n. 2, p. 329-339, 2008.

ROMAN, M.; BRESSAN, D. A.; DURLO, M. A. Morphometric variables and interdimensional relationships for *Cordia trichotoma* (Vell.) Arrâb. Ex Steud. **Ciência Florestal**, v. 19, n. 4, p. 473-480, 2009.

STRADIC, S. L.; BUISSON, E.; FERNANDES, G. W.; MORELLATO, L. P. C. Reproductive phenology of two co-occurring Neotropical mountain grasslands. **Journal of Vegetation Science,** n. 29, p. 15-24, 2018.

SANTOS, E. F.; SALVADOR, T. L.; LIMA-SALVADOR, T.; LEMOS, E. E. P. Physical and chemical characteristics of the fruits of genotypes from the active cambuizeiro germplasm bank. **In: III Symposium of the Northeast Plant Genetic Resources Network**, Aracaju - SE, 2017.

STADNIK, A.; OLIVEIRA, M. I. U.; ROQUE, N. Floristic survey of Myrtaceae in the municipality of Jacobina, Chapada Diamantina, State of Bahia, Brazil. **Hoehnea,** v. 43, n. 1, p. 87,97, 2016.

SEMARH-AL. Municipality of Maceió. **Bulletin December - 2015**. State Secretariat for the Environment and Water Resources. Maceió, AL. January, 2015.

SANTOS, L. C. L.; BEZERRA, Y. C. A.; COSTA, T. D.; REZENDE, L. P.; LEMOS, E. E. P. L. Flowering phenology and fruit biometry of Myrciaria floribunda O. Berg (Myrtaceae) grown in the municipality of Rio Largo - AL. **In: LXV Congresso Nacional de Botànica,** Salvador - Bahia, 2014.

SANQUETTA, C. R.; BEHLING, A.; CORTE, A. P. D.; FERNANDES, A. C.; BECKERT, S. M.; SIMON, A. A. Equations for Estimating Cup Diameter for Black Acacia. **Flora e Ambiente,** p. 14, 2014.

SANTOS, E. D. **Phenology and Fruit Biometry of Cambui (*Myrciaria floribunda* O. Berg.) from Native and Cultivated Populations in Alagoas.** 2010. 75f. Dissertation (Master's Degree in Agronomy: Plant Production) - Federal University of Alagoas (UFAL), Maceió-AL, 2010.

SOUZA, E. B.; NAVES, R. V.; BORGES, J. D.; VERA, R.; FERNANDES, L. P.; SILVA, L. B.; TRINDADE, M. G. Phenology of cagaiteira (*Eugenia dysenterica* DC.) in the State of Goiàs.

Revista Brasileira de Fruticultura, v. 30, n. a, p. 1009-1014, 2008.

SILVEIRA, B. D.; FLORIANO, E. P.; NAKAJIMA, N. Y.; HODOKAWA, R. T.; ROSOT, N. C.; GRACIOLI, C. R. Relationship of morphometry and competition with the growth of *Trichilia claussenii* in a semideciduous forest fragment, RS. **FLORESTA,** v. 45, n. 2, p. 373 - 382, 2015.

SIVIERO, A.; OLIVEIRA, L. C.; PEREIRA, A. A. A.; MENDONÇA, A. S.; NASCIMENTO, F. S. S. ; YUYAMA, K.. Introduction and evaluation of camu-camu (Myrciaria dubia) in Acre. **Proccedings of Congresso Brasileiro de Recursos Genéticos.** Belém, PA; Brazil. Brazilian Society of Genetic Resources, CD ROM, 2012.

TIETBOHl, L. A.C. **Chemical and biological study of the plant species *Myrciaria floribunda* (H.West ex Willd.) O. Berg**. Universidade Federal Fluminense, Niterói-RJ, 155p., Dissertaçao (Mestrado), 2014.

TALORA, D.C.; MORELLATO, P.C. Phenology of tree species in coastal plain forest of southeastern Brazil. **Revista Brasileira Botânica**, v. 23, n. 1, p. 13-26, 2000.

TONINI, H.; ARCO-VERDE, M. F. Canopy morphology to evaluate the living space of four native Amazonian species. **Pesquisa agropecuària brasileira**, v.40, n.7, p.633-638, 2005.

VOGADO, N. O.; CAMARGO, M. G. G.; LOCOSSELLI, G. M.; MORELLATO, L. P. C. Edge effects on the phenology of the guamirim, Myrcia guianensis (Myrtaceae), a cerrado tree, Brazil. **Journal - Tropical Conservation Science**, v. 9, n. 1, p. 291-312, 2016.

ANNEX

Table 6 - Variation and means compared by the Scott-Knott grouping test at 5% probability for the morphological characters of the plants of the *M. floribunda* accessions from the CECA/UFAL Active Germplasm Bank.

Access	Morphological characteristics									
	AP (m)	AIC (m)	AB (m)	DC (m)	DNS (m)	CC (m)	CP (%)	IS	AI	FC
AC02	0,92 p	0,24 s	0,24 x	0,38 z	0,012 m	0,66 m	73,63 f	3,35 i	0,42 h	0,56 m
AC03	1,58 i	1,21 c	0,15 z	1,21 o	0,050 c	1,01 i	62,51 i	2,37 j	0,75 f	1,23 j
AC04	1,47 k	0,48 m	0,30 w	1,16 p	0,034 g	0,91 j	62,02 i	3,39 i	0,81 e	1,28 i
AC05	0,53 s	0,20 t	0,04 β	0,48 x	0,011 m	0,33 p	63,45 i	4,81 g	0,92 e	1,45 g
AC06	1,01 o	0,41 o	0,40 u	1,46 k	0,034 g	0,61 m	58,81 j	4,15 h	1,43 a	2,42 a
AC07	1,32 l	0,59 k	0,50 r	1,21 o	0,056 b	0,69 l	51,81 l	2,15 j	0,91 e	1,75 e
AC08	1,44 k	0,41 o	0,34 v	1,01 r	0,032 h	0,81 k	55,15 k	3,10 i	1,41 a	1,26 i
AC09	1,51 j	0,78 g	0,63 n	1,61 i	0,035 g	0,79 k	53,34 l	4,46 h	0,89 e	2,00 c
AC10	1,26 m	0,54 l	0,41 u	1,11 q	0,033 g	0,91 j	72,01 f	3,24 i	0,67 f	1,23 j
AC11	1,57 i	0,43 n	0,43 t	1,16 p	0,031 h	1,14 h	74,20 e	3,60 i	1,05 d	1,01 k
AC12	1,41 k	0,48 m	0,33 v	1,46 k	0,035 g	1,15 h	82,15 c	4,27 h	0,87 e	1,25 i
AC13	1,03 o	0,61 k	0,36 v	0,97 s	0,020 j	0,41 o	40,01 o	4,54 g	0,75 f	2,36 a

AC14	1,05 o	0,58 k	0,48 s	0,94 t	0,031 h	0,51 n	47,17 m	3,11 i	1,04 d	1,85 d
AC15	1,04 o	0,51 l	0,31 w	1,11 q	0,031 h	0,51 n	64,01 i	3,66 i	0.96 d	2.21 b
AC19	0,86 q	0,13 u	0,14 δ	0,97 s	0,018 k	1,11 h	89,43 b	5,40 f	0,88 e	0,89 l
AC21	1,52 j	0,92 f	0,70 m	1,16 p	0,024 j	0,61 m	72,14 f	4,91 g	1,38 a	1,43 g
AC23	1,22 m	0,64 j	0,30 w	0,97 s	0,021 j	0,89 j	39,23 o	4,12 h	0,79 f	1,92 d
AC24	1,23 m	0,39 p	0,11 *	1,33 m	0,031 h	1,02 i	83,61 c	5,01 g	0,81 e	1,30 i
AC25	1,61 i	0,30 r	0,10 *	1,73 h	0,040 f	1,21 g	74,85 e	4,48 h	0,76 f	1,43 g
AC26	1,11 n	0,47 m	0,31 w	1,13 p	0,022 j	0,55 n	49,10 l	5,39 f	1,09 c	2,10 c
AC27	1,50 j	0,29 r	0,04 β	2,00 d	0,028 i	1,07 i	70,20 g	7,42 c	1,05 d	1,90 d
AC28	1,81 e	0,37 p	0,07 α	2,03 d	0,031 h	1,51 d	82,42 c	6,30 e	1,03 d	1,35 h
AC29	1,36 l	0,52 l	0,05 α	1,03 r	0,016 l	0,71 l	50,70 l	6,12 e	1,32 a	1,49 g
AC31	1,33 l	0,51 l	0,23 x	1,13 p	0,024 j	0,79 k	58,63 j	4,61 g	1,11 c	1,48 g
AC32	1,34 l	0,43 n	0,40 u	1,01 r	0,017 k	0,80 k	60,15 j	5,55 f	0,76 f	1,26 i
AC33	2,03 b	0,52 l	0,50 r	2,00 d	0,040 f	0,91 j	44,56 m	4,86 g	0,87 e	2,23 b
AC34	0,92 p	0,38 p	0,01 β	1,05 r	0,034 g	0,64 m	69,24 g	3,16 i	0,76 f	1,64 f
AC36	0,81 q	0,19 t	0,03 β	1,02 r	0,019 k	0,65 m	80,01 d	5,62 f	0,99 d	1,58 f
AC37	1,01 o	0,42 o	0,12 δ	1,18 o	0,025 i	0,66 m	65,05 h	4,85 g	1,15 c	1,73 e
AC38	1,62 i	0,51 l	0,50 r	1,51 j	0,030 h	1,20 g	75,16 e	5,25 f	1,28 b	1,25 i
AC39	1,44 k	0,43 n	0,48 s	1,33 m	0,028 i	1,06 i	72,42 f	4,98 g	1,14 c	1,28 i
AC40	0,40 t	0,15 u	0,04 β	0,25 δ	0,011 m	0,25 q	61,54 i	3,25 i	0,95 d	1,09 j
AC41	0,91 p	0,20 t	0,03 β	0,37 z	0,009 m	0,75 k	84,10 c	3,01 i	0,93 e	0,49 m
AC42	0,98 p	0,46 m	0,04 β	0,92 t	0,014 l	0,56 n	59,38 j	7,01 c	0,67 f	1,60 f
AC44	1,11 n	0,82 g	0,68 m	0,42 y	0,087 a	0,27 q	29,31 p	0,48 l	0,37 h	1,56 g
AC45	0,86 q	0,53 l	0,35 v	0,73 v	0,040 f	0,39 o	45,07 m	1,82 k	0,84 e	1,87 d
AC46	1,64 h	0,71 e	0,51 r	0,62 w	0,044 d	0,96 j	58,38 j	1,38 k	0,37 h	0,64 m
AC47	1,83 e	0,60 k	0,40 u	1,67 h	0,039 f	1,25 f	68,57 g	4,29 h	0,91 e	1,33 h
AC48	0,91 p	0,36 p	0,25 x	0,63 w	0,019 k	0,45 o	49,97 l	3,36 i	0,69 f	1,40 h
AC49	1,59 i	0,45 n	0,41 u	1,30 m	0,024 i	1,02 i	64,23 i	5,32 f	0,82 e	1,27 i
AC50	1,48 j	0,39 p	0,24 x	1,10 q	0,026 i	1,05 i	72,27 f	4,22 h	0,74 f	1,04 k
AC51	1,64 h	0,46 m	0,20 y	2,01 d	0,042 e	1,22 g	74,68 e	4,79 g	1,22 b	1,64 f
AC52	1,92 d	0,72 h	0,57 p	1,68 h	0,037 g	0,98 j	50,90 l	4,54 g	0,87 e	1,71 e
AC53	0,84 q	0,38 p	0,40 u	0,83 u	0,023 j	0,48 n	56,77 k	3,52 i	0,98 d	1,72 e
AC54	1,81 e	0,53 l	0,31 w	1,85 f	0,042 e	1,35 e	74,58 e	4,41 h	1,02 d	1,37 h
AC55	1,81 e	0,47 m	0,11 *	1,57 i	0,036 g	1,26 f	69,42 g	4,35 h	0,86 e	1,24 i
AC56	1,52 j	0,52 l	0,09 *	1,90 f	0,031 h	1,10 h	72,69 f	6,04 e	1,25 b	1,72 e
AC58	2,15 a	0,79 g	0,44 t	1,08 q	0,026 i	1,21 g	56,50 k	4,07 h	0,58 g	0,88 l
AC59	1,65 h	0,52 l	0,39 u	2,15 c	0,045 d	1,04 i	63,32 i	4,79 g	1,30 a	2,06 c
AC60	1,41 k	0,64 j	0,38 u	1,68 h	0,035 g	0,97 j	68,78 g	4,73 g	1,19 b	1,73 e
AC61	1,61 i	0,46 m	0,23 x	1,75 h	0,036 g	1,37 e	84,83 c	4,81 g	1,08 c	1,28 i
AC62	1,22 m	0,44 n	0,42 t	1,20 o	0,025 i	0,84 k	69,24 g	4,80 g	0,98 d	1,42 g
AC63	1,54 j	0,66 j	0,45 t	1,15 p	0,024 i	1,00 i	64,93 h	4,70 g	0,74 f	1,15 j
AC65	1,08 n	0,53 l	0,35 v	0,98 s	0,028 i	0,65 m	60,37 j	3,46 i	0,90 e	1,50 g
AC66	1,40 k	0,42 n	0,29 w	1,21 o	0,031 h	1,11 h	79,28 d	3,90 h	0,86 e	1,09 j
AC67	1,28 l	0,43 n	0,08 *	1,15 p	0,022 j	0,81 k	63,41 i	5,11 f	0,45 h	1,41 h
AC69	0,92 p	0,19 t	0,25 x	1,17 p	0,028 i	0,72 l	78,80 d	4,11 h	1,27 b	1,61 f
AC70	1,18 m	0,40 o	0,36 v	1,26 n	0,026 i	0,90 j	76,69 e	4,85 g	1,06 c	1,39 h
AC71	1,72 g	0,59 k	0,39 u	1,53 j	0,031 h	1,20 g	69,56 g	4,95 g	0,88 e	1,27 i
AC72	1,64 h	0,34 q	0,11 δ	1,83 g	0,037 g	1,38 e	83,92 c	4,94 g	1,11 c	1,32 h
AC73	1,84 e	0,38 p	0,10 *	1,88 f	0,033 h	1,52 d	82,87 c	5,70 e	1,02 d	1,23 j
AC74	1,97 c	0,70 i	0,43 t	0,99 s	0,030 h	0,77 k	39,08 o	3,24 i	0,50 g	1,28 i
AC75	0,95 p	0,24 s	0,05 α	0,84 u	0,022 j	0,56 n	58,63 j	3,77 i	0,88 e	1,50 g
AC76	1,04 o	0,46 m	0,33 v	0,89 t	0,022 j	0,60 m	56,89 k	4,04 h	0,85 e	1,48 g

AC78	1,76 f	0,30 r	0,06 α	1,69 h	0,027 i	1,34 e	75,93 e	6,29 e	0,95 d	1,26 i
AC79	2,21 a	0,82 g	0,09 *	1,43 k	0,033 h	1,49 d	67,64 h	4,33 h	0,64 f	0,95 k
AC80	1,44 k	0,16 u	0,35 v	1,71 h	0,035 g	1,26 f	87,86 b	4,83 g	1,19 b	1,35 h
AC82	1,01 o	0,24 s	0,38 u	0,86 u	0,023 j	0,85 k	83,74 c	3,73 i	0,84 e	1,01 k
AC84	1,11 n	0,44 n	0,39 u	1,26 n	0,030 h	0,80 k	72,53 f	4,15 h	1,13 c	1,57 g
AC86	1,65 h	0,44 n	1,23 g	1,41 l	0,039 f	1,33 e	80,66 d	3,57 i	0,85 e	1,05 k
AC87	1,59 i	0,40 o	1,26 g	1,59 i	0,034 g	1,40 e	88,04 b	4,67 g	1,00 d	1,13 j
AC88	1,62 i	0,62 k	1,02 j	1,69 h	0,025 i	1,15 h	70,98 f	6,63 d	1,04 d	1,46 g
AC89	1,22 m	0,34 q	0,93 k	1,72 h	0,029 h	0,90 j	73,87 e	6,16 e	1,40 a	1,90 d
AC90	1,56 i	0,53 l	1,02 j	1,88 f	0,044 d	1,32 e	84,61 c	4,22 h	1,20 b	1,42 g
AC91	1,46 k	0,09 v	1,40 e	2,05 d	0,038 f	1,19 g	81,87 d	5,32 f	1,40 a	1,71 e
AC92	1,44 k	0,10 v	1,36 f	1,43 k	0,034 g	1,13 h	78,81 d	4,16 h	0,99 d	1,26 i
AC93	1,45 k	0,38 p	1,14 h	1,63 i	0,025 i	1,17 g	81,02 d	6,54 d	1,12 c	1,39 h
AC94	1,05 o	0,28 r	0,74 l	1,34 m	0,030 h	0,92 j	87,19 b	4,47 h	1,27 b	1,45 g
AC95	1,95 c	0,51 l	1,46 d	2,31 b	0,041 e	1,33 e	68,29 g	5,63 f	1,18 b	1,73 e
AC96	2,18 a	0,23 s	1,98 a	1,76 h	0,043 e	1,48 a	84,62 c	4,05 h	0,80 e	0,95 k
AC97	1,66 h	0,32 q	1,34 f	2,32 b	0,046 d	1,44 e	86,48 b	5,05 g	1,39 a	1,61 f
AC98	1,46 k	0,17 t	1,33 f	1,81 g	0,035 g	1,16 h	79,19 d	5,17 f	1,23 b	1,56 g
AC99	1,19 m	0,59 k	0,63 n	0,82 u	0,015 l	0,62 m	52,50 l	5,53 f	0,69 f	1,32 i
AC100	1,23 m	0,14 u	1,10 i	1,26 n	0,022 j	0,83 k	67,48 h	5,75 e	1,02 d	1,52 g
AC101	1,97 c	0,24 s	1,75 b	1,36 l	0,043 e	1,67 b	84,78 c	4,21 h	0,69 f	0,81 l
AC104	0,68 r	0,19 t	0,53 q	0,61 w	0,014 l	0,58 m	85,32 c	4,21 h	0,89 e	1,05 k
AC105	1,98 c	0,33 q	1,69 c	1,95 e	0,041 e	1,38 e	69,79 g	4,70 g	0,98 d	1,40 h
AC106	2,14 a	0,52 l	0,14 δ	2,49 a	0,043 e	1,78 a	83,17 c	5,79 e	1,16 b	1,39 h
AC107	1,44 k	0,33 q	0,23 x	1,35 m	0,031 h	1,17 g	81,24 d	4,35 h	0,93 e	1,15 j
AC108	0,90 p	0,34 q	0,23 x	0,88 t	0,019 k	0,54 n	59,99 j	4,66 g	0,98 d	1,63 f
AC109	1,97 c	0,30 r	0,46 t	1,71 h	0,043 e	1,61 c	81,77 d	3,99 h	0,86 e	1,06 k
AC110	1,11 n	0,26 s	0,07α	1,01 r	0,040 f	0,86 k	77,92 d	2,52 j	0,90 e	1,16 j
AC111	1,24 m	0,18 t	0,10 *	1,31 m	0,030 h	1,01 i	81,45 d	4,36 h	1,05 c	1,29 i
AC112	1,30 l	0,17 t	0,11 *	1,28 n	0,035 g	1,10 h	84,61 c	3,65 i	0,98 d	1,16 j
AC113	0,96 p	0,19 t	0,18 z	1,14 p	0,026 i	0,84 k	87,57 b	4,41 h	1,18 b	1,35 h
AC114	1,53 j	0,54 l	0,34 v	1,13 p	0,024 i	1,05 i	68,71 g	4,61 g	0,73 f	1,07 k
AC115	2,22 a	0,46 m	0,58 p	1,96 e	0,045 d	1,50 d	67,63 h	4,31 h	0,88 e	1,30 i
AC116	1,49 j	0,36 p	0,44 t	1,66 i	0,030 h	1,19 g	79,59 d	5,45 f	1,11 c	1,39 h
AC117	1,23 m	0,32 q	0,34 v	1,48 j	0,037 g	0,86 k	69,90 g	4,00 h	1,20 b	1,72 e
AC118	1,10 n	0,42 o	0,30 w	1,05 r	0,023 j	0,73 l	66,86 h	4,51 g	0,95 d	1,43 g
AC119	1,22 m	0,18 t	0,34 v	1,13 p	0,026 i	0,98 j	80,36 d	4,28 h	0,93 e	1,15 j
AC120	1,40 k	0,20 t	0,23 x	1,37 l	0,037 g	1,10 h	78,28 d	3,71 i	0,97 d	1,25 i
AC121	1,48 j	0,25 s	0,05 β	1,28 n	0,031 h	1,21 g	81,76 d	4,07 h	0,86 e	1,05 k
AC122	1,82 e	0,82 g	0,65 n	1,52 j	0,037 g	1,03 i	56,71 k	4,10 h	0,83 e	1,46 g
AC123	1,11 n	0,18 t	0,32 w	1,63 i	0,031 h	1,04 i	93,30 a	5,17 f	1,46 a	1,56 g
AC124	1,40 k	0,24 s	0,46 s	1,44 k	0,031 h	1,25 f	89,28 b	4,66 g	1,03 d	1,15 j
AC125	1,69 g	0,23 s	0,32 w	1,70 h	0,044 d	1,36 e	80,23 d	3,87 h	1,00 d	1,25 i
AC126	0,93 p	0,37 p	0,40 u	0,92 t	0,017 k	0,61 m	65,76 h	5,43 f	0,98 d	1,49 g
AC127	0,81 q	0,51 l	0,35 v	0,82 u	0,020 j	0,51 n	62,95 i	4,13 h	1,01 d	1,61 f
AC128	1,57 i	0,64 j	0,51 r	1,39 l	0,035 g	1,00 i	63,06 i	4,18 h	0,89 e	1,45 g
AC129	1,28 l	0,52 l	0,50 r	1,27 n	0,024 j	0,69 l	53,90 k	5,29 f	0,99 d	1,84 d
AC130	2,06 b	0,62 k	0,53 q	1,87 f	0,038 f	1,21 g	58,59 j	4,93 g	0,90 e	1,54 g
AC131	0,97 p	0,41 o	0,38 u	0,71 v	0,019 k	0,41 o	42,54 n	3,76 i	0,73 f	1,72 e
AC132	1,61 i	0,40 o	0,21 y	1,55 j	0,041 e	1,11 h	69,25 g	3,78 i	0,96 d	1,39 h
AC133	1,72 g	0,41 o	0,20 y	1,11 q	0,026 i	1,15 h	67,14 h	4,27 h	0,64 f	0,96 k
AC134	1,11 n	0,58 k	0,35 v	0,75 v	0,019 k	0,51 n	46,38 m	3,98 h	0,68 f	1,46 g

AC135	1,09 n	0,44 n	0,20 y	0,76 v	0,016 l	0,71 l	65,28 h	4,76 g	0,69 f	1,06 k
AC136	1,81 e	0,42 o	0,31 w	1,81 g	0,038 f	1,32 e	73,19 f	4,77 g	1,00 d	1,37 h
AC137	1,91 d	0,61 k	0,55 q	1,62 i	0,031 h	1,01 i	52,87 l	5,22 f	0,84 e	1,60 f
AC138	0,61 r	0,52 l	0,31 w	0,43 y	0,017 k	0,26 q	42,25 n	2,56 j	0,70 f	1,67 e
AC139	1,32 l	0,48 m	0,12 δ	1,41 l	0,026 i	1,03 i	78,01 d	5,32 f	1,06 c	1,37 h
AC140	1,77 f	0,44 n	0,23 x	1,07 q	0,031 h	1,27 f	71,74 f	3,45 i	0,60 g	0,94 k
AC142	1,83 e	0,46 m	0,30 w	1,28 n	0,030 h	1,35 e	74,04 e	4,28 h	0,70 f	0,84 l
AC143	1,81 e	1,02 e	0,65 n	1,82 g	0,037 g	1,20 g	66,11 h	4,92 g	1,00 d	1,51 g
AC144	1,04 o	0,52 l	0,23 x	0,92 t	0,022 j	0.47 o	45.18 m	4,22 h	0,88 e	1,96 c
AC145	1,74 f	0,39 p	0,21 y	1,33 m	0,040 f	1,29 f	73,93 e	3,30 i	0,76 f	1,03 k
AC146	1,55 i	0,41 o	0,51 r	1,83 g	0,041 e	1,17 g	75,82 e	4,41 h	1,18 b	1,55 g
AC147	1,08 n	0,30 r	0,27 x	0,82 u	0,020 j	0,71 l	65,89 h	4,10 h	0,75 f	1,14 j
AC150	1,17 n	0,82 g	0,44 t	1,28 n	0,029 h	1,12 h	95,73 a	4,41 h	1,09 c	1,14 j
AC152	1,54 j	1,33 b	0,19 y	2,10 c	0,038 f	1,28 f	82,84 c	5,46 f	1,36 a	1,64 f
AC153	1,39 k	1,34 b	0,07 α	1,79 g	0,041 e	1,12 h	80,58 d	4,38 h	1,29 b	1,60 f
AC154	1,25 m	0,79 g	0,55 q	1,23 n	0,029 h	1,01 i	80,88 d	4,18 h	0,98 d	1,21 j
AC155	1,27 m	0,74 h	0,53 q	1,04 r	0,027 i	0,96 j	75,61 e	3,78 i	0,81 e	1,08 j
AC156	1,65 h	1,03 e	0,65 n	2,01 d	0,041 e	1,25 f	75,52 e	4,85 g	1,21 b	1,61 f
AC157	1,32 l	1,10 d	0,34 v	1,66 i	0,026 i	1,22 g	92,44 a	6,39 e	1,25 b	1,35 h
AC158	1,42 k	1,03 e	0,39 u	1,87 f	0,032 h	1,11 h	78,16 d	5,88 e	1,32 a	1,68 e
AC159	1,56 i	1,20 c	0,38 u	1,79 g	0,037 g	1,19 g	76,03 e	4,84 g	1,14 c	1,50 g
AC160	1,59 i	1,41 a	0,13 δ	1,74 h	0,014 l	1,04 i	65,40 h	12,49 b	1,09 c	1,67 e
AC161	1,74 f	1,43 a	0,21 y	1,65 i	0,012 m	1,26 f	72,20 f	13,83 a	0,94 e	1,30 i
AC165	1,80 e	1,18 c	0,54 q	1,39 l	0,033 g	1,26 f	70,00 g	4,14 h	0,77 f	1,10 j
AC167	1,22 m	0,94 f	0,43 t	1,55 j	0,038 f	1,17 g	95,50 a	1,96 k	1,26 b	1,32 h
AC170	1,31 l	0,61 k	0,31 w	1,11 q	0,017 k	0,97 j	74,05 e	6,54 d	0,84 e	1,14 j
AC171	0,79 q	0,40 o	0,38 u	0,53 x	0,015 l	0,34 p	43,73 n	3,55 i	0,66 f	1,56 g
AC172	1,40 k	0,59 k	0,09 *	1,44 k	0,035 g	0,95 j	67,96 g	4,07 h	1,02 d	1,51 g
AC173	1,63 h	0,52 l	0,32 w	1,61 i	0,040 f	1,09 h	66,66 h	4,02 h	0,98 d	1,47 g
AC174	1,08 n	0,46 m	0,40 u	0,72 v	0,015 l	0,52 n	47,92 m	4,68 g	0,66 f	1,39 h
AC175	0,57 s	0,30 r	0,22 x	0,56 w	0,015 l	0,41 o	72,80 f	3,67 i	0,99 d	1,36 h
AC177	1,70 g	0,61 k	0,35 v	0,95 s	0,033 g	1,17 g	68,61 g	2,85 i	0,56 g	0,81 l
AC179	1,61 i	0,45 m	0,34 v	1,32 m	0,031 h	1,15 h	71,51 f	4,20 h	0,82 e	1,14 j
AC180	1,60 i	0,73 h	0,51 r	0,95 s	0,026 i	1,11 h	69,68 g	3,67 i	0,59 g	0,85 l
AC181	1,17 n	0,19 t	0,36 v	0,83 u	0,021 j	0,99 i	84,62 c	3,98 h	0,71 f	0,84 l
AC182	1,53 j	0,36 p	0,20 y	1,44 k	0,034 g	1,13 h	73,61 f	4,23 h	0,93 e	1,27 i
AC183	1,03 o	0,04 w	0,47 s	1,08 q	0,026 i	0,96 j	93,25 a	4,07 h	1,04 d	1,11 j
AC184	1,08 n	0,13 u	0,31 w	0,93 t	0,025 i	0,92 j	84,83 c	3,72 i	0,85 e	1,01 k
AC185	1,80 e	0,44 n	0,59 o	1,32 m	0,030 h	1,30 e	72,49 f	4,41 h	0,76 f	1,01 k
AC186	1,65 h	0,50 l	0,54 q	1,36 l	0,039 f	1,20 g	72,72 f	3,50 i	0,82 e	1,13 j
AC187	1,54 j	0,40 o	0,54 q	1,51 j	0,037 g	1,21 g	78,56 d	4,08 h	0,98 d	1,24 i
AC188	1,61 i	0,91 f	0,50 r	1,85 f	0,040 f	1,35 e	84,16 c	4,64 g	1,15 c	1,36 h
AC189	1,51 j	1,23 c	0,30 w	1,70 h	0,030 h	1,23 g	81,18 d	5,57 f	1,12 c	1,38 h
AC190	1,31 l	1,10 d	0,22 x	1,28 n	0,025 i	0,93 j	70,98 f	5,12 f	0,97 d	1,37 h
AC192	1,35 l	0,99 e	0,41 u	1,50 j	0,034 g	1,04 i	77,11 e	4,41 h	1,10 c	1,43 g
AC193	1,36 l	1,14 d	0,22 x	1,68 h	0,031 h	1,17 g	85,71 c	5,35 f	1,23 b	1,44 g
AC194	1,51 j	1,43 a	0,12 δ	2,03 d	0,047 d	1,15 h	75,91 e	4,28 h	1,34 a	1,76 e
AC195	1,76 f	1,33 b	0,53 q	1,56 j	0,031 h	0,97 j	55,11 k	5,03 g	0,88 e	1,60 f
Minimum	0,40	0,04	0,01	0,25	0,008	0,24	24,31	0,48	0,37	0,49
Average	1,40	0,51	0,42	1,33	0,30	0,99	70,21	4,52	0,95	1,38
Maximum	2,22	1,43	1,98	2,50	0,088	1,87	95,73	13,80	1,46	2,42
EP	0,030	0,020	0,020	0,024	0,005	0,017	0,729	0,074	0,012	0,017

CV (%)	2,60	3,09	4,01	2,44	6,01	3,93	2,47	7,95	5,70	4,60

AP - plant height; AIC - crown insertion height; APB - height of the first fork; DC - crown diameter; DNS - stem diameter at 10 cm from the ground; CC - crown length; PC - crown proportion; IS - overhang index; IA - coverage index; FC - crown form; EP - standard error of the mean; CV - coefficient of variation.

Figura 8 - Erect growth habit (A), and more branched, prostrate growth (B), of the *M. floribunda* accessions from the CECA/UFAL Active Germplasm Bank.

PHOTOS: Everton Ferreira dos Santos, 2017.

Figura 9 - Beginning of flowering (A and B), and floral buds of *M. floribunda* accessions from the CECA/UFAL Active Germplasm Bank.

PHOTOS: Everton Ferreira dos Santos, 2017.

Figura 10 - Flower buds (A), flower buds and fruits (B and C), general view of the plant during the fruiting period of *M. floribunda* from the Active Germplasm Bank of CECA/UFAL. PHOTOS: Everton Ferreira dos Santos, 2017.

CHAPTER III

PHYSICAL AND PHYSICAL CHEMICAL CHARACTERIZATION OF FRUITS OF CAMBUI ACCESSORIES (*Myrciaria floribunda* (West ex Willdenow) O. Berg) NATIVE TO ALAGOAS

SUMMARY

The aim of this study was to evaluate the physical and physico-chemical characteristics of the fruits of cambuizeiro accessions native to Alagoas, in order to gather information that will contribute to the domestication and genetic improvement of the species, as well as to identify materials with characteristics that are attractive to the consumer market, thus enabling the establishment of commercial orchards. The fruits used in the research were harvested from the Cambui BAG at CECA/UFAL. The physical and physico-chemical characteristics of the fruit were determined in 11 cambuizeiro accessions. In the physical characterization, 50 measurements were taken individually, consisting of 50 fruits per access evaluated, in terms of the weight of the fruit and seed, in grams; longitudinal and transverse diameter of the fruit, in mm; and diameter of the seed, in mm. The physico-chemical evaluations were carried out using five replicates per access for each characteristic, obtained from 500 g samples of fruit pulp per access. The physico-chemical parameters evaluated were: ST, AT, ST/AT ratio, vitamin C, carotenoids, anthocyanins, flavonoids, total and soluble pectin. The results were subjected to descriptive statistical analysis, obtaining the minimum, mean and maximum values, standard error of the mean and coefficient of variation, and the means were compared using the Scott-Knott grouping test at 5% probability. Pearson's correlation coefficients between the variables were also obtained and verified for significance using the *t-test* ($P>0.01$). There were significant differences between the accessions evaluated in terms of the physical and physico-chemical parameters measured. There is considerable genetic variability in the active cambui germplasm bank, showing that it is a genetically promising population which could serve as a basis for work on the domestication, selection, pre-improvement and improvement of the species.

Index Terms: cambuizeiro; germplasm bank; evaluation; domestication and improvement.

1 - INTRODUCTION

Brazil is considered worldwide to be one of the main chemical, biological and genetic repositories of unexplored or little-known natural technologies, which can be used in various scientific and technological processes, such as in the food, pharmaceutical, cosmetics and fertilizer industries, among others. The country has a great floristic diversity, mainly native fruit trees, which are unknown or little known, with enormous potential for use in the most diverse purposes, many of which have been relegated by Brazilian scientific research. In view of its great continental extension, variety of climates and soils and endemic plant species, it presents a diversity of colors, aromas, flavours and textures of its fruits, which could become attractive to Brazilian and world fruit growing, when they are properly characterized and studied, in the face of the immense market in need of new flavours and products (MUNIZ et al., 2017; SOUZA et al., 2017; SILVA et al., 2008; VIANA et al., 2008; CHITARRA & CHITARRA, 2005).

Therefore, plant genetic resources correspond to a part of biodiversity with a foreseeable current or potential use. These include traditional and improved varieties, advanced lines and native species. With regard to native species, the Brazilian Northeast has two centers of diversity, located in the Caatinga and Atlantic Forest Biomes, where perennial species and native fruit trees stand out. It is estimated that there are around 7,233 accessions, belonging to 38 genera, in fruit germplasm banks in the Northeast, and only around 100 species of native fruit trees occurring in the region have been described and cataloged (JÙNIOR et al., 2014; RAMOS et al., 2008; QUEIROZ et al., 1999). Thus, the state of Alagoas has a very rich and varied flora, with a range of useful plant species as yet unexplored. Botanically, its flora is made up of 150 families, 842 genera and 2,002 species. The Myrtaceae family is considered one of the most representative in terms of richness and diversity (LYRA-LEMOS et al., 2016).

The myrtaceae are one of the most representative botanical families of Brazilian flora, occurring in almost all biomes, and their main characteristic is that they have fruit-producing species with small fruits. Among the species representing this family, *Myrciaria floribunda,* popularly known as "Cambui", stands out. It is a native fruit tree that produces juicy fruits that can reach up to 13 mm in diameter and vary in color from yellow, orange, red and burgundy when ripe. Its fruits are consumed *fresh* or, depending on their physical and chemical characteristics, processed into liqueurs, sweets, juices, jams and ice cream, among others (ARAÙJO, 2012; SANTOS, 2010).

The characterization of plant germplasm banks is an essential stage in genetic improvement programs, as it brings together tools for identifying genotypes and the variability that exists between genotypes, as well as indicating those with potential characteristics for immediate use by farmers, as well as those with peculiar characteristics that are important for the process of improving plant genetic resources (LIMA et al., 2015).

Physical and physicochemical characterization is important for identifying and selecting individuals with potential characteristics for producing fruit for both the fresh and processed fruit markets. It is also a tool for evaluating genetic variability within and between plant populations, allowing species of the same genus to be distinguished, as well as characterizing the correlation between environmental aspects and the genotypic expression of plants (GUEDES et al., 2015; CHITARRA;

ALVES, 2001).

In addition, information on the physicochemical characteristics, nutritional and functional value of fruit is extremely important, as it contributes to an increase in fruit consumption and the development of new products. These attributes are responsible for defining the ways in which these products are consumed and used in the food industry. The richness of vitamins, minerals and fiber among the potential of fruit makes it the most important nutritional food in the population's diet (ZILLO et al., 2014; ROCHA et al., 2013).

The objectives of this study were to evaluate the physical and physicochemical characteristics of the fruits of cambuizeiro accessions native to Alagoas, from the Cambui Active Germplasm Bank (BAG - Cambui) at the Center for Agricultural Sciences of the Federal University of Alagoas (CECA/UFAL), with the aim of gathering information that will contribute to the process of domestication and genetic improvement of the species, as well as identifying materials with characteristics that are attractive to the consumer market.

2 - MATERIALS AND METHODS

Place of execution and collection of plant material

The fruit was collected from the Cambui Active Germplasm Bank (BAG - Cambui), located at the Agricultural Sciences Center of the Federal University of Alagoas (CECA/UFAL), in the municipality of Rio Largo - Alagoas (latitude 9° 29' 45" S, longitude 35° 49' 54" W, altitude 127 meters). According to Koppën, the climate of the region where the germplasm bank is located is classified as tropical rainy, with a dry summer and average annual rainfall of 1,150.2 mm, where November and December are characterized as the driest months of the year and the rainiest from July to August (SEMARH, 2015).

The fruit was harvested when it had reached the ideal point of physiological ripeness, selected and packed in a thermally insulated box with ice, with the aim of maintaining the fruit's physiological characteristics, thus avoiding the loss of moisture and consequent degradation of its chemical constituents. They were then transported to the Fruit Post-Harvest Laboratory at CECA - UFAL for the determination of the fruit's physical and physico-chemical parameters. The fruits were washed in running water and then sanitized in a solution of sodium hypochlorite, in the proportion of 5 mL of chlorine for each liter of water, and then rinsed in running water to remove chlorine residues. The fruit pulp was extracted in a multiprocessor, obtaining ± 500.0 g of pulp from each access, and kept in plastic packaging (153 x 149 x 86 mm) in a freezer at -10°C until the physico-chemical evaluations were carried out.

Physical characterization of the fruit

The color of the fruit from the different accessions was determined using the Munsell Color Charts (1977). The physical variables: fruit weight, seed weight, longitudinal and transverse fruit diameter and seed diameter were determined using the average of 50 fruits from each access. The weight of the fruit and seeds was determined by weighing each fruit and seed individually on a digital scale with 0.01g-Mark 1300 centesimal precision, and the results were expressed in grams (g). The longitudinal diameter (LD) and transverse diameter (TD) of the fruit and seeds were determined using a digital caliper (6G-150 mm), with the results expressed in mm.

Physico-chemical characterization of the fruit

The physico-chemical characteristics, total solids (TS) (°Brix), total acidity (TA), ascorbic acid, carotenoids, anthocyanins, flavonoids, total and soluble pectin, the averages were obtained using five replicates. To assess the quality of the fruit pulp, 1 to 10 g of sample per replication was used for each access.

The Total Solids (TS) content was determined using a digital refractometer model PDR 50B, with a range from 0 to 65°Brix and the results were expressed in °Brix, according to AOAC (1995). Total acidity (expressed as % citric acid) was determined by titration with 0.1N sodium hydroxide solution (NaOH), using phenolphthalein as an indicator (IAL, 2008), in a 25 mL graduated burette, extracting 5 g of pulp and transferring it to a 200 mL conical flask with 50 mL of distilled water. Once the ST and AT values had been determined, the ST/AT ratio was obtained from the quotient of these two variables.

The ascorbic acid content was determined using the amido-iodine method. 10 g of the fruit pulp was weighed into a 100 mL conical flask and then 50 mL of distilled water was added. The sample was homogenized, then 2 mL of the previously prepared solution of sulfuric acid (H_2 SO_4 - 2N), 1 mL of potassium iodide (KI - 10%) and 1 mL of the 10% starch solution were added, the sample was homogenized again. The sample was then titrated in a 25 mL graduated burette with potassium iodate (KIO3 - 0.01 N) until the turning point, when the sample began to turn brown (IAL, 2008). The results were expressed in mg.100g^{-1} .

The total carotenoids present in the fruit of cambuizeiro accessions were quantified according to the AOAC methodology (1995). 5g of the fruit pulp was weighed into a centrifuge tube wrapped in aluminium foil, then 15 mL of isopropyl alcohol (P.A.) and 5 mL of hexane (P.A.) were added, and the sample was shaken for 2 minutes in a homogenizer. After stirring, the contents were transferred to a 125 mL separating funnel wrapped in aluminum foil, where the volume was topped up with distilled water and the sample was left to stand for 30 minutes, then washed (removing the aqueous phase and leaving the yellow phase). After three rests of 30 minutes each, the sample was filtered through cotton wool (sprayed with a pinch of anhydrous sodium sulphate P.A.) so that it could be dried.

a 25 mL volumetric flask wrapped in aluminum foil, washing the cotton with hexane and pressing it in the funnel to remove all the pigment present in the cotton. After filtering, 2.5 mL of acetone was added to the sample and the remaining volume was completed with hexane. A solution consisting of 5 mL of acetone and 45 mL of hexane was used as a blank, and the readings were taken on a spectrophotometer at a wavelength of 450 nm. Total carotenoids were calculated using the formula described below:

$$\text{Total carotenoids (mg\%)} = \frac{\text{ABS x 100}}{\text{250 x 1 X weight/dilution}}$$

Where,

ABS - absorbance obtained when read on a spectrophotometer;

Weight - amount of pulp used in the analysis;.

Dilution - final volume of the carotenoid extract obtained.

Total anthocyanins and total flavonoids were determined according to the methodology developed by Francis (1982). 5 g of fruit pulp was weighed into a beaker wrapped in aluminum foil, then 30 mL of the previously prepared ethanol (95%) / HCl (1.5 N) extraction solution was added in the ratio 85:15. The samples were homogenized in a Turrax tissue homogenizer for 2 minutes at speed "5". The contents were then transferred to a 50 mL volumetric flask (unfiltered) in a dark place, filled with the extraction solution, homogenized and stored in an amber bottle, which was left to stand overnight in the fridge. The following day, the material was filtered in a 100 mL kitassato coupled to a vacuum pump protected from light. Readings were taken on a spectrophotometer at a wavelength of 535 nm for anthocyanins and 374 nm for flavonoids. The anthocyanin and flavonoid contents were calculated using the formula: absorbance x dilution factor/98.2, and the results were expressed in $mg.100^{-1}$.

The total pectic substances were extracted and quantified according to the methodology developed by McReady and MacComb (1952). 5 g of pulp was weighed in a Becker, then 25 mL of EtOH (95%) was added, the sample was homogenized and left to stand for 30 minutes in the fridge. After resting, the sample was filtered through a 100 mL kitassato coupled to a vacuum pump, and the residue was washed twice with 75% ethanol. The residue was then transferred to a 50 mL beaker, where 40 mL of distilled water was added, homogenizing the sample again. After this, the pH of the sample was adjusted to 11.5 with a 1 N NaOH solution and then left to stand for 30 minutes, again in the refrigerator. The pH was then adjusted to 5.5 with diluted glacial acetic acid (5 mL of CH_3 COOH/ 50 mL of distilled water), to provide ideal reaction conditions in the sample, after the inoculation of pectinase from *Aspergillus niger* v. Tiegem. After extraction, readings were taken by spectrophotometry at 520 nm. The percentage of total pectin was calculated using the weight of the sample, the absorbance obtained and the equation resulting from the calibration curve (x = y - 0.0064/0.00421), where the y in the equation was replaced by the absorbance of the sample, obtaining the pectin content in micrograms (μg), which was then transformed into a percentage using a simple rule of three.

The soluble pectic substances were extracted and quantified according to the methodology developed by McReady and MacComb (1952). 5 g of pulp was weighed in a Becker, then 25 mL of EtOH (95%) was added, the sample was homogenized and left to stand for 30 minutes in the fridge. After resting, the sample was filtered through a 100 mL kitassato coupled to a vacuum pump, and the residue was washed twice with 75% ethanol. The residue was then transferred to a 100 mL erlenmeyer flask, 40 mL of distilled water was added and the sample was stirred for 1 hour. The material was then vacuum filtered and the supernatant diluted to 100 mL in a volumetric flask. After extraction, readings were taken by spectrophotometry at 520 nm. For the quantification of soluble pectin, the same procedure was used as for total pectin, i.e. the percentage of total pectin was calculated using the weight of the sample, the absorbance obtained and the equation resulting from the calibration curve (x = y - 0,0064/0.00421), where the y of the equation was replaced by the absorbance of the sample, obtaining the pectin content in micrograms (μg), which was then transformed into a percentage using a simple rule of three.

Statistical Analysis

The data obtained was subjected to the Kolmogorov-Smirnov and Lilliefors normality tests to verify the normality and homogeneity of the variances. Subsequently, the data was subjected to descriptive statistical analysis, obtaining the maximum, mean and minimum values and the standard error of the mean. Pearson's correlation coefficient between the variables was calculated, and significance was verified using the *t-test* ($P>0.01$). To assess the genetic variability between the accessions, based on the standardized means of the characteristics studied, the means were compared using the Scott-Knott grouping test at a 5% probability level. All statistical analyses were carried out using the GENES software (CRUZ, 2006).

3 - RESULTS AND DISCUSSION

The fruits of *M. floribunda* varied in color from orange to purple (Table 1), emanating a citrus aroma with a sour and slightly sweet taste, persistent and very pleasant to smell. Knowledge of the physical characteristics of the fruit, *in terms* of external appearance, size, shape and color of the skin, are key quality attributes for marketing the fruit *in natura* and using the pulp to make industrialized products, as well as providing important data for handling and storing the fruit, as well as helping to standardize laboratory tests and develop techniques for storing seeds and producing seedlings (SILVA et al., 2013; ALVES et al., 2012).

Table 1 - Fruit color of *M. floribunda* accessions, native to Alagoas, collected from the Cambuizeiro Active Germplasm Bank of the Agricultural Sciences Center of the Federal University of Alagoas (CECA/UFAL), located in Rio Largo - Alagoas, 2018. Adapted from KAISER (2016).

Access	Epicarp color	Letter from Munsell
AC04	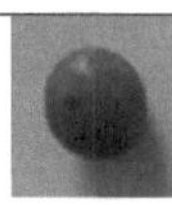	7.5 YR 7/8
AC05		5 YR 6/8
AC29		7,5 R 3/1
AC32		10 R 5/8
AC38		5 YR 8/8
AC39		7,5 R 3/2
AC50		10 R 5/8
AC71		5 YR 6/8
AC128	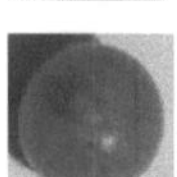	7.5 YR 7/8
AC152		5 YR 6/8

AC192		5 YR 6/8

There were significant differences between the accessions for all the physical characteristics evaluated (Table 2). It can therefore be seen that different groupings were formed, expressing the potential for the plant population to distinguish between the characters measured, establishing genetic differences in the population of cambui accessions native to Alagoas.

Four groups (a to d) were formed for the fruit weight parameter; five groups (a to e) were obtained for seed weight; four groups (a to d) for the longitudinal diameter of the fruit and three groups for the transverse diameter (a to c); three groups (a to c) were obtained for the DL/DT ratio and three groups (a to c) were also formed for seed diameter. Therefore, fruit weight, seed weight and longitudinal diameter can be considered superior characters for expressing the genetic diversity existing in the active cambuizeiro germplasm bank.

The weight of the fruit varied between 0.71 and 1.26 g, with the orange-colored accessions AC50, AC152 and AC192 having the highest average weights. The lowest values were observed in accessions AC04, AC05 and AC38, with average weights of 0.72, 0.72 and 0.71 g, respectively. The average weight of the fruit is a valuable characteristic, given that the fruits with the highest weight are usually also the largest, making them more attractive to the consumer (LIMA et al., 2002). The results obtained in this study are superior to those reported by Santos et al. (2014), who, when evaluating the physical characteristics of 25 individuals of *M. floribunda* grown in Rio Largo - Alagoas, observed fruit with an overall average of 0.73 ± 0.11 g. Another study by Araùjo (2012) found average weights ranging from 0.552 ± 0.104 g to 1.083 ± 0.151 g in fruits from native populations of cambui from the coastal vegetation of Alagoas.

Table 2 - Physical characteristics of the fruits of cambuizeiro (*Myrciaria floribunda* O. Berg) accessions native to Alagoas, from the CECA/UFAL Active Germplasm Bank.

Access	**Parameters**					
	FP (g)	**PS (g)**	**DL (mm)**	**DT(mm)**	**DL/DT**	**DS (mm)**
AC04	0,72 d	0,18 d	8,71 c	9,01 b	0,96 b	4,68 c
AC05	0,72 d	0,17 d	8,30 c	8,88 b	0,93 c	5,88 b
AC29	0,94 b	0,44 a	8,88 c	8,95 b	0,98 a	5,48 c
AC32	0,55 e	0,10 e	8,36 c	8,38 c	0,99 a	4,75 c
AC38	0,71 d	0,17 d	8,20 c	9,11 b	0,89 c	4,68 c

AC39	0,81 c	0,25 c	8,50 c	9,27 b	0,91 c	5,26 c
AC50	1,26 a	0,44 a	10,17 a	11,03 a	0,91 c	6,07 b
AC71	0,49 e	0,12 e	7,30 d	7,88 c	0,92 c	6,80 b
AC128	0,86 c	0,20 c	9,32 b	9,71 b	0,95 b	11,05 a
AC152	1,16 a	0,31 b	9,46 b	10,71 a	0,89 c	11,90 a
AC192	1,17 a	0,25 c	10,34 a	11,05 a	0,92 c	5,89 b
Overall average	0,85	0,24	8,86	9,45	0,93	6,59
CV (%)	11,40	19,16	5,51	6,23	2,72	11,43

***Means** followed by equal letters belong to the same group according to the Scott & Knott test at 5% probability; PF - Fruit Weight; PS - Seed Weight; LD - Longitudinal Diameter; TD - Transverse Diameter; LD/DT - relationship between the variables; SD - Seed Diameter.

With regard to seed weight, there was also heterogeneity between the accessions, with values ranging from 0.17 to 0.44 g, with accessions AC29 (0.44 g), AC50 (0.44 g) and AC151 (0.31 g) showing the highest values. Intermediate weights were shown by accessions AC39, AC128 and AC192, with averages of 0.25, 0.20 and 0.25 g, respectively. Accessions AC04 (0.18 g), AC05 (0.17 g) and AC38 (0.17 g) had the lowest values. Santos (2010), when characterizing cambuizeiro fruits from native and cultivated populations in Alagoas, found an average seed weight of 1.10 g, differing from the results obtained in this research.

Seed diameter, which had an overall average of 6.59 mm, varied from a minimum of 4.68 mm and 5.48 mm in accessions AC04, AC38 and AC29 to a maximum of 11.90 mm in accession AC152. The biometric characterization of seeds is related to the dispersal and establishment characteristics of seedlings, and is also used as a tool to distinguish pioneer and non-pioneer plant species in tropical forests (MONTEIRO et al., 2016).

On the other hand, the coefficients of variation of the dimensional characteristics, such as longitudinal diameter (5.51 %), transverse diameter (6.23 %) and the ratio between the DL/DT variables (2.72 %), show that the data analyzed for these variables was less dispersed, in other words, more homogeneous, given that the coefficients of variation for fruit weight, seed weight and seed diameter were greater than 11.40%, indicating a more heterogeneous distribution between the accessions.

The average LD values ranged from 8.20 (AC38) mm to 10.34 (AC192) mm between the accessions, with an average of 8.86 mm. With regard to DT, the ranges observed were from 7.88 (AC71) mm to 11.05 (AC192) mm, with a mean of 9.45 mm. The DL and DT results presented in this study corroborate those obtained by Araujo (2012), who, when assessing the physical characteristics of the fruit of cambui genotypes native to Alagoas, observed DL dimensions ranging from 9.30 mm to 12.18 mm for red and yellow genotypes, and transverse diameters ranging from 8.39 mm to 10.75 for red and purple genotypes, respectively.

The ratio between the length and width (DL/DT) of the fruit is used as an indication of its shape. Table 2 shows that the fruit of the cambuizeiro accessions had a general average of 0.93 for this characteristic, with a range of values between 0.89 (AC152) and 0.99 (AC32), i.e. DL DT values ≤ 1, indicating slightly rounded or oblong fruit, which is a very important characteristic, especially for fruit intended for processing, such as fruit used in the manufacture of preserved jams, where the final appearance of the product is essential for its acceptance by the consumer. Furthermore, processing industries prefer rounder fruit, as it makes cleaning and processing easier (SILVA et al., 2013; CHITARRA & CHITARRA, 2005).

The variability observed in the physical characteristics of the fruit of the different cambuizeiro accessions stems from the fact that they originate from different mother plants, in addition to the segregation provided by allogamy, which generally occurs in fruit species. However, such variability becomes a tool of exceptional importance, as it enables the selection of plant materials with superior genetic characteristics, as well as due to its usefulness in the conservation of germplasm and the collection of propagation material (BORGES et al., 2010).

Table 3 shows the analytical results of the physicochemical characterization of the fruits of 11 cambuizeiro accessions. Based on the statistical analysis of the 9 physicochemical parameters determined, it can be seen that different groupings were also formed between the accessions, demonstrating the ability of certain characteristics to establish the genetic diversity existing between the accessions. Thus, for ST, four groups were obtained (a to d); for AT, five groups were established (a to e); the ST/AT ratio also showed five groups (a to e); for vitamin C content, three groups were determined (a to c); for total carotenoids, the accessions were subdivided into ten groups (a to j); for anthocyanins, four groups were formed (a to d); seven clusters were formed for flavonoids (a to g); the total pectin content was grouped into five categories among the accessions (a to f); and four clusters were established for total pectin (a to d).

The total solids (TS) content ranged from 11.50°Brix for access A192 to 18.78°Brix for access A05, with an average value of 14.01°Brix and a coefficient of variation of 7.06%. According to Correa et al. (2011), this parameter is used as a quality attribute related to taste, indirectly indicating the concentration of sugar dissolved in the pulp or juice of the fruit, not indicating the appropriate levels, given that other chemical compounds are also dissolved, and is also used as an indication of the degree of maturity for some fruits. In this way, it can be seen that in the physico-chemical characterization of the fruits of the different cambui accessions, considerable amounts of organic acids, vitamin C, carotenoids, anthocyanins, flavonoids and pectic substances were detected, which may explain the wide ranges observed for total solids between the accessions.

In the physico-chemical characterization of the fruit of cambui genotypes carried out by Araujo (2012), total solids contents ranging from 8.96°Brix to 11.46°Brix were obtained. Pinheiro et al. (2011) studied the genetic diversity of a natural population of *Myrciaria tenella* O. Berg and found TS levels ranging from 10.75°Brix to 16.31°Brix. It can be seen that the cambui accessions evaluated in this study had higher ST levels than those reported by the aforementioned authors.

Table 3 - Physico-chemical characteristics of the fruits of cambuizeiro (*Myrciaria floribunda* O. Berg) accessions native to Alagoas, from the CECA/UFAL Active Germplasm Bank.

Access	Parameters								
	ST	AT	ST/AT	VC	CT	AN	FL	EN	PS
AC04	14,58 c	4,49 d	3,24 c	49,63 c	2,35 a	4,43 c	4,62 g	0,06 d	0,24 c
AC05	18,78 a	7,15 b	2,64 d	58,08 c	1,72 d	4,48 c	8,18 f	0,03 e	0,09 d
AC29	11,88 d	2,41 e	4,96 a	103,48 a	1,18 h	22,29 b	23,63 b	0,35 a	0,38 a
AC32	17,92 a	4,69 d	3,86 b	55,96 c	1,23 g	2,60 d	18,73 c	0,13 c	0,19 c
AC38	13,22 d	5,37 c	2,46 d	40,12 c	1,65 e	33,56 a	24,48 b	0,30 b	0,38 a
AC39	16,36 b	4,92 d	3,35 c	48,22 c	0,31 j	4,20 c	14,93 d	0,13 c	0,21 c
AC50	12,92 b	7,99 a	1,61 e	64,41 c	2,17 b	2,38 d	7,54 f	0,05 e	0,21 c
AC71	12,38 d	8,10 a	1,54 e	42,24 c	2,11 c	2,53 d	8,71 f	0,07 d	0,05 d
AC128	12,48 d	5,79 c	2,15 d	48,57 c	2,21 b	2,11 d	25,71 a	0,08 d	0,29 b
AC152	12,12 d	8,42 a	1,45 e	44,35 c	1,38 f	2,70 d	2,03 h	0,07 d	0,21 c
AC192	11,50 d	3,07 e	3,76 b	76,03 b	1,08 i	1,76 d	13,34 e	0,09 d	0,07 d
Average	14,01	5,67	2,82	57,37	1,58	7,55	13,81	0,12	0,21
CV (%)	7,06	9,43	12,00	29,03	2,05	10,50	5,82	18,20	18,88

*Means followed by equal letters belong to the same group according to the Scott & Knott test at 5% probability. ST: Total Solids (°Brix); TA: Total Acidity (%); ST/AT: ratio between the two variables; VC: Vitamin C ($mg.100g^{-1}$); CT: Total Carotenoids (mg%); AN: Anthocyanins ($mg.100g^{-1}$); FL: Flavonoids ($mg.100g^{-1}$); PT: Total Pectin (%); PS: Soluble Pectin (%).

The fruits of the cambui tree are acidic, with total acidity (TA) values ranging from 2.41% to 8.42%

equivalent to citric acid, in accessions A29 and A152, respectively, with an average of 5.67%, higher than those presented by Araùjo (2012) and Santos (2010), who observed an average value of 1.40% in the fruits of genotypes from native and cultivated populations of cambui in Alagoas. In the work carried out by Santos et al. (2014), in which they assessed the genetic diversity between accessions of araça (*Psidium* spp.) from different municipalities in the semi-arid region of Bahia, they detected total acidity ranging from 1.01 to 9.62%, values very close to those found in this study for *M. floribunda fruit.*

In this way, the characterization of total acidity is essential in foodstuffs, as it helps to determine their useful post-harvest life. According to Lima et al. (2002), high acidity (>1%) is a peculiar characteristic for fruit destined for processing, as it reduces the addition of synthetic acidifiers to the processed product, which is a technique used in the food industry to inhibit the proliferation of microorganisms, thus prolonging the shelf life of the food. It is known that the percentage of organic acids present in fruit varies greatly depending on the fruit species, as well as due to the use of acids by the plant itself in the Krebs cycle and/or due to their transformation into sugars during the respiration process. The concentration of these compounds decreases as the fruit ripens, and can be used as a reference for the degree of ripeness and taste of the fruit, whether sweet or sour (BUENO et al., 2017). Therefore, the results obtained for TA indicate that the fruit has characteristics that can be used both for *fresh* consumption and for use in food processing industries.

The ratio between total solids and total acidity (ST/AT) or ratio was quite variable between the accessions, with a coefficient of variation of 12 %. The range of variation was from 1.45 for access A152 to 4.96 for access A29, with a midpoint of 2.82. Silva et al. (2017) found a range of 0.13 to 3.17 in camucamuzeiro clones from the Active Germplasm Bank of Embrapa Amazônia Oriental. In *Myrciaria jabuticaba* (Vell) Berg. fruit from different producing regions in the state of São Paulo, the ST/AT ratio ranged from 7.43 to 18.98 (OLIVEIRA et al., 2003).

The ST/AT ratio is one of the best ways of evaluating fruit flavor, representing an indication of the sweet-sour nature of the fruit, i.e. the balance between acids and sugars, where its value increases as the fruit passes from its "once" to "ripe" stage, Furthermore, this ratio is used to determine the degree of sweetness of the fruit, especially in the selection of raw materials used in agro-industrial processing, and is an indicator of the need to correct °Brix or acidity (GADELHA et al., 2009). According to Camilo et al. (2014), fruits with a high ST/AT ratio are more acceptable to the population for *fresh* and/or processed consumption, as they are sweeter and less acidic. Thus, accessions A29, A32 and A192 would be the most suitable, with averages of 4.96, 3.86 and 3.76, respectively.

The vitamin C content detected in the cambui fruit ranged from 40.12 $mg.100g^{-1}$ (A38) to 103.48 $mg.100g^{-1}$ (A29), with an average of 57.37 $mg.100g^{-1}$. The coefficient of variation was 29.03 %, indicating the high variability between the accessions in this area. It is worth noting that among the fruit species belonging to the Myrtaceae family, camu-camu (*Myrciaria dùbia* (H. B. K.) McVaugh) is one of the largest sources of vitamin C. Thus, Smiderle et al. (2008), when quantifying vitamin C in camucamuzeiro fruit, obtained a range of 2.59 to 2.52 g per 100g of pulp.

In cambuizeiro genotypes from the coastal vegetation of Alagoas, values ranging from 517.80 ± 128.12 $mg.100g^{-1}$ to 1481.81 ± 128.33 $mg.100g^{-}$ 1 were found, well above the levels obtained in this study (ARAÙJO, 2012). However, the values presented in this study can be considered high given that the recommended daily intake of vitamin C for adults is 60 mg (BRASIL, 2005). According to Valillo et al. (2005), the content of these compounds varies depending on the genetic characteristics of the plant, environmental factors, the degree of ripeness and the post-harvest handling and storage of the fruit. It can therefore be concluded that these factors may explain the variations observed in vitamin C content between the accessions of *M. floribunda* characterized.

With regard to the concentration of total carotenoids, the accessions showed levels ranging from 0.31 mg% (AC39) to 2.35 mg% (AC04), with an average point of 1.58 mg% and a coefficient of variation of 2.05%, indicating low variability between the accessions in this regard. In quantifying the total carotenoid content in fruits of araçâ-amarelo, araçâ-vermelho (*Psidium cattleyanum* Sabine) and araçà-pera (*P. acutangulum* D.C.) grown in Pelotas/RS, Fetter et al. (2010) found values of 1.07 ± 0.08 $mg.100g^{-1}$, 0.99 ± 0.16 $mg.100g^{\wedge 1}$ and 0.59 ± 0.09 $mg.100g^{\wedge 1}$, respectively. In the chemical characterization of the fruits of *M. floribunda* collected in sandbank areas of Maricâ/RJ by Oliveira et al. (2018), an average total carotenoid content of 52.22 ± 0.54 $mg.100g^{-1}$, i.e. the results obtained in this study were lower. However, the cambuizeiro germplasm collection has potential characteristics for selection in genetic improvement programs, with the aim of obtaining fruits with high carotenoid contents.

With regard to the content of total anthocyanins, the highest content was obtained by access A38 with 33.56 $mg.100g^{-1}$, while the lowest value was observed in access AC192 with 1.76 $mg.100g^{-1}$, with intense orange and red colored fruit, respectively. The general average among the accessions was 7.55 $mg.100g^{-1}$, with a coefficient of variation of 10.50%. As for the quantitative determination of total flavonoids, the highest content was found in access AC128 with 25.71 $mg.100g^{-1}$ and the lowest in access AC152 with 2.03 $mg.100g^{-1}$. The average content observed was 13.38 $mg.100g^{-1}$, with a coefficient of variation of

5,82 %. According to the coefficients obtained, there is reasonable variability in the cambuizeiro population for these characteristics.

In the characterization of wild fruits from southern Brazil, the total anthocyanin content of araçà (*P. cattleianum*) fruits was 2.98 ± 0.58 mg.100g^{-1} (Fritsch et al., 2014). Vizzotto et al. (2012), when characterizing the fruit of *P. cattleianum,* found an average content of 38.3 ± 12.0 mg.100g^{-1} . In quantifying pigments during the development of camu-camu (*M. dùbia*) fruit, Silva et al. (2012) detected amounts ranging from 7.37 mg.100g^{-1} to 170.00 mg.100g^{-1} , during the course of development until complete physiological ripeness.

In the study carried out by Lima et al. (2011), who quantified the content of total anthocyanins in the fruit of two varieties of jaboticaba (*Myrciaria cauliflora (*Mart.) O. Berg), they found that the highest concentrations of these chemical compounds were found in the epicarp, with the Paulista variety having a content 22.95% higher than the Sabarà variety, with a content of 383.0 mg.100g^^{1} and 362.0 mg.100g^{-1} , for the two varieties respectively.

The quantification of phenolic compounds in ripe *Myrcia multiflora* fruit carried out by Pinto et al. (2017) found a yellow flavonoid content of 2858.50 mg.100g^{-1} . Grigio et al. (2017), when quantifying the concentration of flavonoids in different parts of the fruit of camu-camu (*M. dubia*), obtained values of 440.9 ± 4.9 µg.g^{-1} , 91.5 ± 6.5 µg.g^{-1} , 550.8 ± 12.3 µg.g^{-1} , 517.6 ± 21.0 µg.g^{-1} and 49.6 ± 2.8 µg.g^{-1} , in the peel and pulp, pulp, peel, whole fruit and seed, respectively. Virgolin et al. (2017) observed an average content of 1.56 ± 0.03 mg.100g^{-1} of pulp in the fruit of araçâ-boi (*Eugenia stipitata*). Araujo et al. (2015) carried out a chemical characterization of myrtle fruit (*Eugenia gracillima* Kiaersk.) and found a content of 69.53 ± 0.36 mg.100g^{-1} .

By comparing the concentration of anthocyanins and flavonoids found in the fruit of the different cambui accessions characterized in this study with the results obtained by the aforementioned authors for other fruit species of the Myrtacea family, it can be seen that the content of these chemical compounds varies greatly between species and parts of the fruit. The content of bioactive compounds in fruit pulp is influenced by numerous factors, such as ripening stage, soil and climatic conditions, geographical location of the orchard, mineral nutrition, ripening stage, post-harvest handling and, above all, the variability existing in the plant's genetic material (CHITARRA; CHITARRA, 2005).

With regard to total pectin content, using the Scott-Knott cluster analysis, the accessions were grouped into five groups, where group "c" grouped 45.45% of the accessions (AC04, AC71, AC128, AC152 and AC192). The highest average total pectin content was observed in AC29, with 0.35%. The lowest content was found in access AC05, with an average of 0.03%. The overall average obtained was 0.12% and the coefficient of variation was 18.20%, showing considerable genetic divergence between the accessions. For soluble pectin, there was considerable variability between the accessions, with values ranging from 0.05% to 0.38%, with accessions AC29 (0.38%)

and AC38 (0.38%) showing the highest percentages. The lowest content was obtained in access AC71 (0.05%), and the overall average was 0.21%, with a coefficient of variation of 18.88%.

In the study of the physical, chemical and physicochemical characterization of myrtle fruits (*Eugenia gracillima* Kiaersk.) carried out by Araujo et al. (2015), average total pectin contents of 612.66 ± 5.95 mg.100g^^{1} and soluble pectin of 116.88 ± 6.09 mg.100g^^{1} were obtained. Pereira et al. (2015) quantified the pectin content in the fruit of *Syzygium paniculatum* Gaertn. and *Syzygium cumini* (L.) Skeels, commonly known as jamelao, cherry, olive, jalao, myrtle, among other names, and found soluble pectin content in the fruit pulp ranging from 128.05 ± 0.08 mg.100g^{-1} to 189.81 ± 16.68 mg.100g^{-1} , total pectin ranging from 54.18 ± 3.06 mg.100g^{-1} to 47.66 ± 0.47 mg.100g^{-1} , for *S. paniculatum and S. cumini,* respectively. Yuyama et al. (2002) quantified the content of soluble and insoluble dietary fiber present in the fruit pulp in different camu-camu populations and found values of 0.99 ± 0.18 mg.100g^^{1} to 3.55 ± 0.01 mg.100g^^{1} for insoluble fiber and 0.51 ± 0.01 mg.100g^{-1} and 0.84 ± 0.01 mg.100g^^{1} for soluble fiber. Thus, the total and soluble pectic substance contents obtained from the cambuizeiro fruit were lower than those reported by the authors mentioned above for other fruit trees in the Myrtaceae family.

However, the cambui accessions showed considerable levels of pectic substances, indicating their potential for agro-industrial processing, in the production of jams, jellies, compotes, among others, due to their ability to form gels, in addition to contributing to the promotion of human health. Given that these compounds have beneficial effects on health, preventing and/or reducing cardiovascular diseases, gastrointestinal problems, cancer, sugar and blood pressure control, obesity, among other benefits (CHITARRA & CHITARRA, 2005).

Table 3 shows the Pearson correlation coefficients (r) of all the physical parameters of the fruit of the cambuizeiro accessions evaluated. According to Lima et al. (2015), evaluating the correlations between characters of interest is extremely important, as it reveals in advance the influence of selecting a specific character on the change in the average value of the other.

Table 4 - Pearson's correlation coefficients between the six physical characteristics evaluated in the fruits of the cambuizeiro accessions native to Alagoas.

Parameters	PF	PS	DL	DT	DL/DT	DS
PF	---	0,780 ^{ns}	0,913 ^{ns}	0,938 ^{ns}	-0,255 *	0,340 *
PS		---	0,605 ^{ns}	0,568 ^{ns}	-0,016 ^{ns}	0,119 ^{ns}
DL			---	0,938 ^{ns}	-0,034 ^{ns}	0,267 *
DT				---	-0,362 **	0,344 **

DL/DT	---	-0,196 ns
DS		---

ns, **, * Non-significant and significant at 1 and 5% probability levels using the t-test, respectively; PF: Fruit Weight; PS: Seed Weight; LD: Longitudinal Diameter; TD: Transverse Diameter; LD/DT: ratio between the two variables; SD: Seed Diameter.

Most of the correlations were not significant, but there was a positive and significant correlation between fruit weight (PF) and fruit transverse diameter (DT) and seed diameter (DS), indicating that the greater the PF and DT, the greater the DS. Significant and negative correlations were also observed between the weight and transverse diameter of the fruit and the DL/DT ratio, so we can infer that the lower the weight and transverse diameter, the lower the ratio between the length and width of the fruit.

Simple correlations are usually used in studies of long-cycle plants, especially native species, with the aim of understanding the relationships between the characteristics evaluated, as well as being used as a selection strategy, with a view to increasing genetic gains indirectly (LIMA et al., 2014).

As shown in Table 5, positive correlations were observed between total acidity (TA) and total carotenoids (r = 0.413); and between total acidity and anthocyanins (r = 0.332), indicating that total acidity in fruit increases as the content of these chemical compounds increases. There was a negative and significant correlation between total solids (TS) and total pectin (TP), predicting that the latter tends to decrease as the concentration of total solids in the fruit pulp increases. The correlation coefficient between the ST/AT ratio and the characters anthocyanins (AN), flavonoids (FL) and soluble pectin (PS) were r = 0.265, r = 0.428 and r = 0.233, showing that the ST/AT ratio is linearly proportional to the AN, FL and PS content.

As for the vitamin C (VC) content, there is a positive linear correlation with the total pectin content. However, the lower the percentage of total pectin in the fruit, the lower the concentration of total carotenoids, as they showed a negative and significant correlation (r = -0.336). Highly significant correlations were not detected for all the characters, however, the results obtained in this work indicate the feasibility of selecting accessions with superior characteristics by analyzing the correlated parameters.

Table 5 - Pearson's correlation coefficients between the nine physical and chemical characteristics evaluated in the fruits of the accessions of *M. floribunda* native to Alagoas.

Parameters	ST	AT	ST/AT	VC	CT	AN	FL	EN	PS
ST	---	0,049 ns	0,206 ns	-0,152 ns	-0,191 ns	-0,168 ns	-0,086 ns	-0,247 *	-0,195 ns

AT	---	-0,910 ns	0,520 ns	0,413 **	0,332 *	-0,554 ns	-0,561 ns	-0,320 *
ST/AT		---	0,585 ns	-0,496 **	0,265 *	0,428 **	0,519 ns	0,233 *
VC			---	-0,192 ns	0,119 ns	0,203 ns	0,310 **	0,101 ns
CT				---	-0,095 ns	-0,202 ns	-0,336 **	-0,016 ns
AN					---	0,563 ns	0,870 ns	0,682 ns
FL						---	0,680 ns	0,568 ns
EN							---	0,710 ns
PS								---

ns, **, * Not significant and significant at 1 and 5% probability level by t-test, respectively; ST: Total Solids; TA: Total Acidity; ST/AT: ratio between the two variables; VC: Vitamin C; CT: Total Carotenoids; AN: Anthocyanins; FL: Flavonoids; PT: Total Pectin; PS: Soluble Pectin.

4 - CONCLUSIONS

1 - The fruit has physical characteristics that can be used in food processing industries to produce jams, jellies, ice creams, compotes, drinks and more;

2 - The cambui accessions evaluated show a considerable degree of variation in terms of the fruit's physicochemical characteristics, which are favorable both for *fresh* consumption and for processing, making it possible to select genotypes based on a single superior trait or concomitantly with other traits;

3 -The cambui fruits evaluated have great potential for industrialization due to their quality characteristics, such as high juiciness, high total solids content, high acidity and high vitamin C concentration;

4 - The fruits of the cambuizeiro accessions are potential sources of compounds with antioxidant properties, such as anthocyanins, flavonoids and carotenoids, which could make it easier to introduce them to the market as a differentiated product;

REFERENCES

ARAÙJO, D. R.; LUCENA, E. M. P.; GOMES, J. P.; FIGUEIRÊDO, R. M. F.; SILVA, E. E. Physical, chemical and physico-chemical characteristics of myrtle fruit. **Revista Verde de Agroecologia e Desenvolvimento Sustentâvel,** v. 10, n. 3, p. 11-17, 2015.

ARAÙJO, R. R. **Quality and utilization potential of fruit from Cambui, Guajiru and maçaranduba genotypes native to the coastal vegetation of Alagoas**, Mossoró, 175p., Tese (Doutorado), 2012.

ALVES, A. M.; FERNANDES, D. C.; SOUSA, A. G. O.; NAVES, M. M. V. Physical and chemical characterization of pequi fruits from three Brazilian states. **Ciência Rural,** Santa Maria, 2012.

AOAC - Association of official analytical chemists international. **Official Methods of Analysis.** 16 ed. Arlington, 1995. v. 2, 474p.

BUENO, G. H.; GUEDES, M. N. S.; SOUZA, A. G.; MADEIRA, A. P. C.; GARCIA, E. M.; TAROCO, H. A.; MELO, J. O. F. Physical and physicochemical characterization of *Eugenia dysenterica* DC. fruits from a high-altitude tropical climate region. **Revista Brasileira de Biomedicina,** v. 35, n. 3, p. 515-522, 2017.

BORGES, K. C. F.; SANTANA, D. G.; MELO, B.; SANTOS, C. M. Pulp yield and fruit and seed morphometry of pitangueira-do-cerrado. **Revista Brasileira de Fruticultura,** v. 32, n. 2, p. 471-478, 2010.

BRAZIL. ANVISA. Resolution RDC No. 269, of September 22, 2005. **Approves the "Technical Regulation on the Recommended Daily Intake of Proteins, Vitamins and Minerals".** D. O. U. - Official Gazette of the Union. ANVISA - National Health Surveillance Agency, of September 23, 2005.

CAMILO, Y. M. V.; SOUZA, E. R. B.; VERA, R.; NAVES, R. V. Fruit characterization and selection of cagaite tree (E *ugenia dysenterica* DC.) progenies. **Cientifica**, v. 42, n. 1, p. 1-10, 2014.

CORREIA, A. A. S.; GONZAGA, M. L. C.; AQUINO, A. C.; SOUZA, P. H. M.; FIGUEIREDO, R. W.; MAIA, G. A. Chemical and physico-chemical characterization of noni pulp (*Morinda citrifolia*) grown in the state of Cearà. **Alimentos e Nutriçâo**, v. 22, p. 609615, 2011.

CRUZ, C.D. **GENES program**: multivariate analysis and simulation. Viçosa: UFV, 2006. 175p.

CHITARRA, M. I. F.; CHITARRA, A. B. **Post-harvest of fruits and vegetables: physiology and handling.** Lavras: UFLA, 785 p., 2005.

CHITARRA, A. B.; ALVES, R. E. **Post-harvest Technology for Tropical Fruits**. Fortaleza: Instituto de Desenvolvimento da Fruticultura e Agroindùstria - FRUTAL/ Sindicato dos Produtores de Frutas do Estado do Cearà, 436 p., 2001.

FRITSCH, M.; BOHN, A.; KUHN, F.; SCAPINELLO, J.; DAL MAGRO, J. Characterization of the antioxidant activity of wild fruits from the Southern Region of Brazil. **In: XX Congresso Brasileiro de Engenharia Quimica,** 7 p., Florianópolis - SC, 2014.

FETTER, M. R.; VIZZOTTO, M.; CORBELINI, D. D.; GONZALEZ, T. N. Functional properties of araçâ-amarelo, araçâ-vermelho (*Psidium cattleyanum* Sabine) and araçâ-pera (*P. acutangulum* D.C.) grown in Pelotas/RS. **Brazilian Journal of Food Technology,** III SSA, 4 p., 2010.

FRANCIS, F.J. Analysis of anthocyanins. In: MARKAKIS, P. (ed.). **Anthocyanins as food colors.** New York: Academic Press, 1982. p.181-207.

GRIGIO, M. L.; CHAGAS, E. A.; RATHINASABAPATHI, B.; CHAGAS, P. C.; SILVA, A. R. V.; SOBRAL, S. T. M.; OLIVEIRA, R. R. Qualitative evaluation and biocompounds present in different parts of camu-camu (*Myrciaria dubia*) fruit. **African Journal of Food Science,** v. 11, n. 5, p. 124-129, 2017.

GUEDES, M. L.; FERREIRA, P. H. G.; SANTANA, K. N. O.; PIMENTA, M. A. S.; RIBEIRO, L. M. Fruit Morphology and Productivity of Babassu Palms in Northern Minas Gerais State, Brazil. **Revista Arvore**, v. 39, n. 5, p. 883-892, 2015.

GADELHA, A. J. F.; ROCHA, C. O.; VIEIRA, F. F.; RIBEIRO, G. N. Avaliação de parâmetros de qualidade fisico-quimicos de pulpas congeladas de abacaxi, acerola, cajà e caju. **Revista Caatinga,** v. 22, n. 1, p. 115-118, 2009.

IAL - INSTITUTO ADOLFO LUTZ. **Physico-chemical Methods for Food Analysis.** 4ª ed., 1° ed. Digital, v. 1, Sao Paulo - SP, 1020 p., 2008.

JÙNIOR, J. F. S.; LEDÓ, A. S.; SILVA, A. V.C.; RAMOS, S. R. R. Genetic Resources of Native and Adapted Fruit Trees in the Northeast: status of germplasm conserved *ex situ* in the region. In: III Brazilian Congress on Genetic Resources, 2014, Santos - SP. **Proceedings...** 4 p., 2014.

KAISER, D. K. **Physiological maturity, desiccation tolerance and longevity of *Allophylus edulis* [(A. ST. - HIL., A. JUSS. & CAMBESS.) HIERON. EX NIEDERL] seeds.** Dissertation (Master's Degree in Agronomy) - State University of Western Paranà, Campus Marechal Càndido Rondon, 89 f., 2016.

LYRA-LEMOS, R. P.; MOTA, M. C. S.; SILVA, J. W. A.; SILVA, A. C. M.; PRATA, A. P. N. Flora of Alagoas: state of the art, advances and future perspectives. In: 67th National Botanical Congress, 2016, Vitória - ES. **Proceedings...** 1 p., 2016.

LIMA, M. S. S.; DANTAS, A. C. V. L.; FONSECA, A. A. O.; BARROSO, J. P. Fruit Characterization of Selected Genotypes of Umbu-cajazeira (*Spondias* sp.). **Interciencia,** v. 40, n. 5, p. 311-316, 2015.

LIMA, C. A.; FALEIRO, F. G.; JUNQUEIRA, M. T. V.; BELLON, G. Evaluation of physicochemical characteristics of fruits from two pitaya species. **Revista Ceres,** v. 61, n. 3, p. 377-383, 2014.

LIMA, A. J. B.; CORRÊA, A. D.; SACZK, A. A.; MARTINS, M. P. CASTILHO, R, O. Anthocyanins, pigment stability and antioxidant activity in jabuticaba [*Myrciaria cauliflora* (mart.) o. berg]. **Revista Brasileira de Fruticultura,** v. 33, n. 3, p. 877- 887, 2011.

LIMA, M. A. C.; ASSIS, J. S.; NETO, L. G. Characterization of guava fruit and selection of cultivars in the Sao Francisco Submedium Region. **Revista Brasileira de Fruticultura,** v. 24, n. 1, p. 273-276, 2002.

MUNIUZ, J.; PELIZZA, T. R.; LIMA, A. P. F.; GONÇALVES, M. J.; RUFATO, L. Postharvest Quality of Red Araçà. **Revista U.D.C.A Actualidad & Divulgación Cientifica,** v. 20, n. 2, p. 311-319, 2017.

MONTEIRO, R. A.; FIOREZE, S. L.; NOVAES, M. A. G. Genetic variability of *Erythrina speciosa* matrices based on morphological characters. **Sciencia Agraria Paranaensis (Online),** v. 15, n. 1, p. 48-55, 2016.

MUNSELL COLORS CHARTS. **Munsell colors charts for plant tissues**. 2ª ed. New York, 1977.

McCread, P. M,;McComb, E. A.; Extraction and determination of total pectin materials. **Analytical Chemistry**, Washington, v. 24, n. 12, p. 1586-1588, 1952.

OLIVEIRA, L. M.; PORTE, A.; GODOY, R. L. O.; SOUZA, M. C.; PACHECO, S.; SANTIAGO, M. C. P. A.; GOUVÊA, A. C. M. S.; NASCIMENTO, L. S. M.; BORGUINI, R. G. Chemical characterization of *Myrciaria floribunda* (H. West ex Willd) fruit. **Food Chemistry,** n. 248, p. 247-252, 2018.

OLIVEIRA, A. L.; BRUNINI, M. A.; SALANDINI, C. A. R.; BAZZO, F. R. technological characterization of 'Sabarâ' jabuticabas from different growing regions. **Revista Brasileira de Fruticultura,** v. 25, n. 3, p. 397- 400, 2003.

PINTO, M. K. N. A.; NAZARENO, L. S. Q.; RUFINO, M. S. M.; LOPES, M. M. A.; MIRANDA, M. R. A. Quantification of phenolic compounds and total antioxidant activity of ripe cambui (*Myrcia multiflora).* **In: XXXVI Congresso de Iniciaçâo Cientifica,** Encontros Universitàrios da UFC, Fortaleza - CE, v. 2, 2017.

PEREIRA, R. J.; CARDOSO, M. G.; VILAS BOAS, E. V. B.; PEREIRA, R. J. Quality aspects and centesimal composition of the fruits of *Syzygiuui cumini* (L.) Skeels and *Syzygiuui paniculatum* Gaertn. **Revista Cereus**, v. 7, n. 1, 18 p., 2015.

PINHEIRO, L. R.; ALMEIDA, C. S.; SILVA, A. V. C. Genetic diversity of a natural population of cambuizeiro and post-harvest evaluation of its fruit. **Scientia Plena,** v. 7, n. 6, 5 p., 2011.

QUEIROZ, M. A.; GOEDERT, C. O.; RAMOS, S. R. R. **Recursos Genéticos e Melhoramento de**

Plantas para o Nordeste Brasileiro. (on line). Version 1. 0. Petrolina - PE: Embrapa Semiàrido/ Brasilia - DF: Embrapa Recursos Genéticos e Biotecnologia, 1067 p., 1999.

ROCHA, M. S.; FIGUEIREDO, R. W.; ARAÙJO, M. A. M.; MOREIRA-ARAÙJO, R. S. R. Physico-chemical characterization and antioxidant activity (*in vitro*) of fruits from the cerrado of Piauí. **Revista Brasileira de Fruticultura**, Jaboticabal - Sao Paulo, v. 35, n. 4, p. 933941, 2013.

RAMOS. S. R. R.; QUEIROZ, M. A. de; ROMÂO, R. L.; SILVA JUNIOR, J. F. da. Plant germplasm conserved in the Brazilian Northeast: current situation, priorities and prospects. **Magistra**, Cruz das Almas, BA, v. 20, n. 3, p. 205-217, jul./set., 2008.

SOUSA, I. J. O.; ARAÙJO, S.; NEGREIROS, P. S.; FRANÇA, A. R. S.; ROSA, G. S.; NEGREIROS, F. S.; GOLÇALVES, R. L. G. The Diversity of Brazilian Flora in the Development of Health Resources. **Revista UNINGA**, v. 31, n. 1, p. 35-39, 2017.

SILVA, A. T. L.; SILVA, J. C. O.; NASCIMENTO, W. M. O.; SILVA, A. R. L. Physicochemical characterization in fruits of camucamuzeiro clones. **In: 57th Brazilian Congress of Chemistry,** Gramado - Rio Grande do Sul, 2017.

SEMARH-AL. Municipality of Maceió. **Bulletin December - 2015**. State Secretariat for the Environment and Water Resources. Maceió, AL. January, 2015.

SANTOS, M. A. C.; QUEIROZ, M. A.; SANTOS, A. S.; SANTOS, L. C.; CARNEIRO, P. C. S. Genetic diversity among araçà accessions from different municipalities in the semi-arid region of Bahia. **Revista caatinga,** v. 27, n. 2, p. 48, 57, 2014.

SANTOS, L. L. C.; BEZERRA, Y. C. A.; COSTA, T. D.; REZENDE, L. P.; LEMOS, E. E. P. Flowering Phenology and Fruit Biometry of *Myrciaria floribunda* O. Berg (Myrtaceae) Cultivated in the Municipality of Rio Largo-AL. **In: LXV Congresso Nacional de Botànica**, Salvador-Bahia, 2014.

SANTOS, E. D. **Phenology and Fruit Biometry of Cambui (*Myrciaria floribunda* O. Berg.) from Native and Cultivated Populations in Alagoas.** 2010. 75f. Dissertation (Master's Degree in Agronomy: Plant Production) - Federal University of Alagoas (UFAL), Maceió-AL, 2010.

SILVA, L. R.; BARRETO, N. D. S.; MENDONÇA, V.; BRAGA, T. R. Physical and physico-chemical characteristics of water from green dwarf coconut palm fruits. **Revista Brasileira de Tecnologia Agroindustrial,** v. 7, n. 2, p. 1022-1032, 2013.

SILVA, V. X.; PONTIS, J.; FLACH, A.; LIMA, C. G. B.; CHAGAS, E. A.; NEVES, L. C. Quantitative pigment composition during camu-camu (*Myrciaria dùbia*) fruit development. **In: XXII Congresso Brasileiro de Fruticultura,** 4 p., Bento Gonçalves - RS, 2012.

SILVA, G. G.; SOUZA, P. A.; MORAIS, P. L. D.; SANTOS, E. C.; MOURA, R. D.; MENEZES, J. B. Characterization of Wild Plum Fruits (*Ximenia americana* L.). **Revista Brasileira de Fruticultura**, v. 30, n. 2, p. 311-314, 2008.

SMIDERLE, O. J.; SOUSA, R. C. P. Vitamin C content and physical characteristics of camu- camu at two stages of ripeness. **Revista Agro@mbiente On-line,** v. 2, n. 2, p. 61-63, 2008.

VIRGOLIN, L. B.; SEIXAS, F. R. F.; JANZANTTI, N. S. Composition, content of bioactive compounds, and antioxidant activity of fruit pulps from the Brazilian Amazon biome. **Pesquisa Agropecuària Brasileira**, v. 52, n. 10, p. 933-941, 2017.

VIZZOTTO, M.; BIALVES, T. S.; ARAÙJO, V. F.; NACHTIGAL, J. C. Fruit pulps: sources of antioxidant compounds. **In: 4th Food Safety Symposium,** 3 p., Gramado - RS, 2012.

VIANA, B. F.; OLIVEIRA, D.; NASCIMENTO, I. A.; MATUTE, R. G.; BENCHIMOL, R.

L. **Biodiversity and its applications: partnerships between Brazil and Canada, results and future prospects.** Interfaces Brasil/Canadà, n. 9, 14 p., 2008.

VALLILO, M. I.; GARBELOTTI, M. L.; OLIVEIRA, E.; LAMARDO, L. C. A. Physical and chemical characteristics of cambucizeiro (*Campomanesia phaea*) fruits. **Revista Brasileira de Fruticultura**, v. 27, n. 2, p. 241- 244, 2005.

YUYAMA, L. K. O.; AGUIAR, J. P. L.; YUYAMA, K.; LOPES, T. M.; FAVARO, D. I. T.; BERGL, P. C. P.; VASCONCELOS, M. B. A. Quantification of dietary fiber in some populations of cubiu (*Solanum sessiiflorum* Dunal), camu-camu (*Myrciaria dùbia (*H. B. K.) Mc Vaugh) and açai (*Euterpe oleraceae* Mart.). **Acta Amazônica,** Manaus, v. 32, n. 3, p. 491-497, 2002.

FSC
www.fsc.org
MIX
Papier aus verantwortungsvollen Quellen
Paper from responsible sources
FSC® C105338

Printed by Books on Demand GmbH, Norderstedt / Germany